THE CANAL AGE

Also available in the David & Charles Series

THE SOMERSET & DORSET RAILWAY – *Robin Atthill*
BUYING ANTIQUES – *A. W. Coysh and J. King*
INTRODUCTION TO INN SIGNS – *Eric R. Delderfield*
THE SAILOR'S WORLD – *Captain T. A. Hampton*
OLD DEVON – *W. G. Hoskins*
LNER STEAM – *O. S. Nock*
RAILWAY ADVENTURE – *L. T. C. Rolt*
THE WEST HIGHLAND RAILWAY – *John Thomas*
BRITISH STEAM SINCE 1900 – *W. A. Tuplin*

The David & Charles Series

THE CANAL AGE

CHARLES HADFIELD

UNABRIDGED

PAN BOOKS LTD : LONDON

First published 1968 by David & Charles (Publishers) Ltd.
This edition published 1971 by Pan Books Ltd,
33 Tothill Street, London, S.W.1.

ISBN 0 330 02678 X

Printed in Great Britain by
Richard Clay (The Chaucer Press), Ltd., Bungay, Suffolk

To the Boatmen of
THE CANAL AGE

CONTENTS

ILLUSTRATIONS IN PHOTOGRAVURE

LIST OF MAPS

PREFACE

In their time, canals played their part in building Britain. Without inland water transport between 1760 and 1850, industry and agriculture would have developed differently, and more slowly. In this country, though not on the Continent or in the United States, only a few waterways are now used for carrying goods. But many have found a hopeful future in pleasure cruising, and enter their third century with confidence in what is to come and with pride in the past.

The man in the motor cruiser or on the towpath will often wonder how the waterways came to be, and what the canal age was like. I hope this book will tell him some of the things he wants to know. I have added short chapters on the great ship canals, and on the age on the Continent and in North America, so that readers may see a little of the wider background against which the affairs of Britain were set. A bibliography suggests further reading on a subject that does not lose its interest with familiarity.

I have travelled waterways at home and abroad; have been a canal enthusiast for as long as I can remember; and spent four fruitful years as a member of the British Waterways Board. I am also fortunate enough to have lived in the canal age itself, as I have daily felt and realized it in many hundreds of minute books, reports, letters, maps and newspapers of the time. It was not Arcady; far from it. But it was exciting and successful, and now is little known.

CHARLES HADFIELD

ACKNOWLEDGEMENTS

I should like to thank Mr David St J. Thomas, who suggested that I should write this book, and encouraged me while doing so.

I should also like to thank the following for their special help: Mr G. J. Biddle, Mr J. H. Boyes, Mr L. A. Edwards, Mr J. F. Goodchild, Mr C. N. Hadlow, Dr J. R. Hollick, Mr G. Ottley, Dr Roger Pilkington, Mr Harold Sumption, M. R. Thélu, Mr John Thomas and Mr F. Underwood. Also Miss M. M. Davis of the Newcomen Society, the Director of the Historical Society of Western Pennsylvania, Mrs Gail M. Gibson of the Pennsylvania Historical & Museum Commission, Miss Annamary Spanger of the Historical Society of York County, Pennsylvania, and Mrs Stuart Gibson of the Valentine Museum, Richmond, Va, USA.

Mr R. J. Dean has kindly drawn the maps and the cover picture is reproduced by courtesy of the Museum of British Transport and the Hugh McKnight Photography Collection.

CHAPTER 1

How Canals Began

THIS is a book about canals in the age before railways and roads took a large share in heavy goods carrying. But because canals are still an important part of the scheme of things, the story has been brought down to our own times. Again, it is mainly about the short and brilliant canal age in the British Isles, but it says something about matching periods elsewhere.

First comes the natural river. If it runs quickly enough and is sufficiently deep, cargo-carrying craft can sail or be poled along it. If a towing path has been built alongside, horses can tow barges. Nowadays self-propelled diesel craft can navigate it, as on the Meuse, or tugs pushing their barges in front, as on the Mississippi. But most rivers vary greatly in their depth and speed of current at different times of the year: after the wet season they will run fast; in dry periods there may be only a few inches of water. This was the case with our own Severn and Trent in the eighteenth century and earlier. So for ease of navigation, weirs were built across the rivers, and locks inserted in them for the barges to pass through.

A weir holds back the water in a stretch of river, and so slows the current. Instead of the water level falling gradually, the fall is concentrated at the points where it foams over the weir. Boats, however, must now be deliberately lowered from one level to the next.

In the old days – in this country the eighteenth century and earlier – weirs were often built not to help navigation, but to hold back water to provide power for corn, cloth or other mills on the river banks. Those were days when industry depended much more upon water than on steam power. Locks were primitive then. One sort was the flash-lock. A movable section was built into the weir so that by first removing vertical planks or paddles, and then the horizontal beams that held them in place, an opening was made. When this was done, water poured through the gap, lowering the river level above and raising it below.

Barges would shoot down through the gap, while those going upstream would be winched through the opening with a pulley and line from the bank, or would have to wait until the level had equalized. Some flash-locks remained on the upper Thames until this century. Other ways of doing the same thing were the vertically-rising gate or staunch, many of which used to exist on Fenland waterways such as the River Lark, and the single pair of sideways-moving gates, that could be hauled open by a capstan and rope against the pull of the current. These were called water gates or half-locks. Two of them, three hundred years old, were removed a few years ago when the Lower Avon was restored between Evesham and Tewkesbury.

Such contraptions had a life span of over two thousand years, for they are known on Chinese waterways in the last half of the century before Christ, especially in the form of staunches with vertically-rising gates. They were so wasteful of water, however, that Chinese engineers then built slipways up which boats could be dragged from one level to the other. But dragging caused damage, and damage was an incentive to theft of the cargoes. China, therefore, moved another step, by building a chamber in the weir, with sets of gates at each end. A boat going upstream then entered the empty chamber. After the lower gates had been shut, water could be admitted through openings in the upper gates or the sides of the chamber until it was level with the section of river above the weir. Then the upper gates could be opened for the boat to proceed. So the weir itself could be undisturbed, and the only water used would be that which passed through the chamber.

As far as we know, the invention of the ordinary or pound-lock is to be attributed to Chhiao Wei-Yo, assistant commissioner for transport on a section of the Grand Canal of China in the year AD 983, who created it as a solution to slipway problems, and the first, 250 ft long, was built on the West river near Huai-yin. It had vertically-rising gates of a type seen in Britain on the River Nene, and on the Continent on the upper Rhine locks, the Amsterdam–Rhine Canal and elsewhere.

On the Continent of Europe, staunches or flash-locks are known to have existed in Holland in 1065, in Flanders in 1116 and in Italy in 1198. The first pound-lock known dates from 1373, at Vreeswijk in Holland on the River Lek, still an important canal town. This was, however, more of a basin connecting river and canal where boats could wait, than a true lock, such as that known in 1396 at Damme near Bruges. All these, and later examples, had vertically-rising gates. It was Leonardo da Vinci, painter and much else, who in his capacity as engineer to the Duke of Milan invented and built examples of locks with swinging or mitre-gates, of the kind that can be seen on any British canal today.

Another step in improving a river for navigation was to straighten it by building artificial cuts or canals across the bends. This process, which could also include deepening and providing locks, might result in more new cuts than remaining portions of the old river. Such drastically reconstructed river navigations in England were, or are, the Wey, Kennet, Itchen and Welland. An Ordnance map will show what was done.

From river straightening it was an easy step to building a true canal to by-pass a section of difficult river. This was taken for the first time in Britain between 1564 and 1566, when Exeter Corporation, with John Trew as engineer, built the first Exeter Canal, 1¾ miles long and 16 ft wide, from Exeter to below Countess Wear. On it were Britain's first pound-locks, still with vertically-rising gates. A few years later the first mitre-gate lock was probably built at Waltham Abbey on the River Lee.

> This locke contains two double doores of wood,
> Within the same a cesterne all of Plancke,
> Which onely fills when boats come there to passe
> By opening of these mighty dores.

Another remarkable early example is the eight-mile long canal, with twelve locks, built before 1670 to by-pass the section of the River Welland between Market Deeping and Stamford. The biggest canal in Britain, and one of the biggest in the world, the Manchester Ship, is an artificial by-pass to the navigable rivers Mersey and Irwell, leading to Manchester. The St Lawrence Seaway, mainly in Canada, is a series of such by-passes.

An early step was to build a canal extension to, or branch from, a river, such as the Droitwich Canal from the Severn, on which salt could be shipped out from and coal brought into the town. A more difficult one was to connect two water heads by an artificial cut between them. For the first of these we have to go back again to China to the 'Magic Canal', built over a saddle in the hills of north-east Kuangsi province in Shih Lu in 219 BC.

On the Continent the Canal du Midi is a remarkable watershed cut. It was built in the reign of Louis XIV between 1666 and 1681 to link the Mediterranean with the Alantic by a heavily-locked waterway from the River Garonne near Toulouse to the Etang de Thau near Sète, 150 miles long, 64 ft wide, with three important aqueducts. The engineer John Smeaton called it 'the noblest work of the kind that has ever been executed'. Another was the Ludwigs Kanal, 112 miles long and with 100 locks, built between 1836 and 1845 by King Ludwig of Bavaria from Bamberg on the River Main to Dietfurt on the Altmühl to connect the Rhine and Danube. In England the first was Brindley's Trent & Mersey Canal, 93 miles long and with 74 locks, from the River Trent above Nottingham to Preston Brook, where it joined the Duke of Bridgewater's Canal to the Mersey at Runcorn. The most notable we have are perhaps the Leeds & Liverpool between the Mersey and the Aire over the Pennines; the Kennet & Avon, linking the Thames, by way of the Kennet, with the Bristol Avon, and the Forth & Clyde across the waist of Scotland.

There are some magnificent examples across the Atlantic – the New York State Barge, formerly the Erie Canal, from the Hudson river at Albany to Lake Erie at Buffalo; the Rideau Canal in Canada from the river at Ottawa to Kingston on Lake Ontario; and the longest canal ever built, the former Wabash & Erie, from Evansville on the Ohio river to Toledo on Lake Erie. In Ireland the Grand Canal from the sea at Dublin to the Shannon is of this type.

There is a kind of canal built primarily to drain the land, but which may be useful also for navigation. The Romans built our own Fossdyke from Lincoln to the Trent as a drainage canal, but they used it, as we do, for navigation; Vermuyden cut the Dutch river to divert the River Don, and the New Bedford river to supplement the Ouse, to lessen flooding, but made both navigable.

Finally, there are canals that owe little or nothing to rivers – such as the Göta, running from lake to lake across Sweden from Gothenburg to Soderköping south of Stockholm, the Mittelland Canal that sweeps west to east across Germany, or our own Grand Union from London to Birmingham. These lead one to the sea-to-sea waterways, the greatest products of the canal builders' art: our own Caledonian, the Kiel, Corinth, Suez and Panama.

The canal age in the world, which contributed greatly to building its industry and trade, and made possible the kind of industrial revolution we had in Britain, perhaps began when Ptolemy II (285–246 BC) of Egypt finished a waterway between the Nile and the Red Sea that the Pharaoh Necho had begun about 600 BC, and Darius, King of Persia, had continued. In Holland, Flanders, Germany and Italy it began in the Middle Ages; in France perhaps with the Canal de Briare, completed in 1642; in the United States not until the Santee in South Carolina was finished in 1800. But in the British Isles we may date it from March 1742, when the Newry navigation in the north of Ireland was opened by the Lough Neagh colliers *Boulter* and *Cope*, laden with Tyrone coal for Dublin.

CHAPTER 2

They Come to Britain

IN 1729 the Irish Parliament set up Commissioners of Inland Navigation for Ireland, in general to promote waterways, and in particular to build a canal between the coal-mining area of Tyrone south-west of Lough Neagh and the port of Newry, whence coal could be shipped to Dublin to compete successfully, the members hoped, with that from Britain.

Their first engineer was Richard Castle, probably a French refugee Huguenot who had fled to Germany, and had then travelled on the Continent studying navigation works. During his three years' work on the Newry Canal he built what was perhaps the first stone pound-lock in Ireland. He was then dismissed and returned to his other profession of architect in Dublin. He was succeeded by Thomas Steers, one of the first Englishmen to make a full-time profession of civil engineering.

Steers had been born in 1672, and when eighteen he joined the Army, went to Ireland and fought at the Battle of the Boyne. Then he was moved to Holland for four years, where he must have seen many canals and harbour works before returning to London, probably to work on the Howland dock. Dock engineering there and in Liverpool led him to take part in canalizing the Mersey & Irwell rivers from Warrington to Manchester and the River Douglas to Wigan. In 1736 he was asked to take on the Newry Canal, and over the next five years he gave 25 months of his time to it. Some of the rest went to being mayor of Liverpool in 1739.

Eighteen miles long, about 45 ft wide, 5 to 6 ft deep and with 14 locks to take 50-ton barges, the Newry was a major canal,

1 England, Wales and southern Scotland: inland waterways in 1760

2 England, Wales and southern Scotland: inland waterways in 1790

and although Steers' work was to be much criticized later, it served for 60 years. Rebuilt in the first decade of the 1800s, it was successful and useful until, almost 200 years after Steers built it, the last barge passed in 1930. Canals have a long life compared to most other forms of transport.

Steers died in 1750, and was succeeded as Liverpool dock engineer by Henry Berry, aged probably 31, who had been born at Parr near St Helens and started his working life by repairing local roads before becoming Steers' assistant. He must have heard about the work on the Newry Canal, as well as about dock engineering, from his master, for the Liverpool Common Council chose him and William Taylor to survey the unimpressive Sankey Brook in 1754 to see whether this small stream running from near St Helens to the River Mersey could be made a means of bringing coal down from the local mines.

Having been born near it, he must have known that it was too small. A canal was the obvious answer. But a few months before, Parliament had rejected a project only a few miles away, for a canal from Salford through Leigh to Wigan, which had also been surveyed by William Taylor, and might do the same for this one. So, when the proprietors obtained their enabling Act in 1755, it was 'for making navigable the River or Brook called Sankey Brook', but it included the usual power to make 'such new cuts . . . as they shall think proper and requisite'. It seems then that, working quietly with John Ashton, also a Parr man, who held 51 out of 120 shares in the company, he got on with building a canal that was to be nearly 12 miles long, with 12 locks and three branches, using the Brook only for water supply and for a short section at its lower end. The main line and one branch were open in 1757, a second branch by 1762, the third by 1773. The Sankey Canal began to carry large quantities of coal across the Mersey and up the navigable River Weaver to the works around Northwich and Winsford, where salt was obtained from brine. Coal also went down the Mersey to other works, one owned by John Ashton, where rock salt also mined beside the Weaver was refined. Thereafter the development of the St Helens coalfield chiefly depended upon how near the mines were to the canal. As the company paid an average dividend of 33⅓ per cent for the next eighty years, the other shareholders probably did not complain seriously of Berry's and Ashton's deception.

A few miles north-west of the Sankey Canal lay Worsley Hall, a manor-house owned by the Duke of Bridgewater, not far from his collieries. There in 1757 came the young Duke, then aged twenty-one. At seventeen he had gone on the Grand Tour, and during it had visited and interested himself in the Canal du Midi, and also taken a course in experimental philosophy, which would have included some science and engineering. Back in London, he enjoyed himself for a time before deciding to visit one of his estates.

He came to Worsley Hall and there met John Gilbert, the agent in charge of his lands and collieries, then aged thirty-three. Soon they were deep in plans, centred on the problem of extending the market for their coal, a market that indeed had just begun to contract as that from the St Helens district carried cheaply down the Sankey began to replace some of that brought by waggon from Worsley.

There were several factors. In 1737 a group of Manchester men, supporters of the Mersey & Irwell river navigation, had got an Act to make the Worsley Brook navigable to the Irwell near Barton. It was not followed up, perhaps because two of the leading promoters had died. But quickly-growing Salford and Manchester needed coal. In 1753 an Act was passed to improve the local roads from Salford, including that from Worsley; and in 1754 came the abortive scheme already mentioned, for a canal from Salford to Wigan, which also was probably indirectly promoted by Mersey & Irwell interests.

The Duke started by asking the Mersey & Irwell company how much they would charge for his traffic up the Irwell from Barton to Manchester if he got it that far. It is alleged they replied that there would be no reduction on the toll from Warrington. This ruled out a canal to the Irwell, and suggested

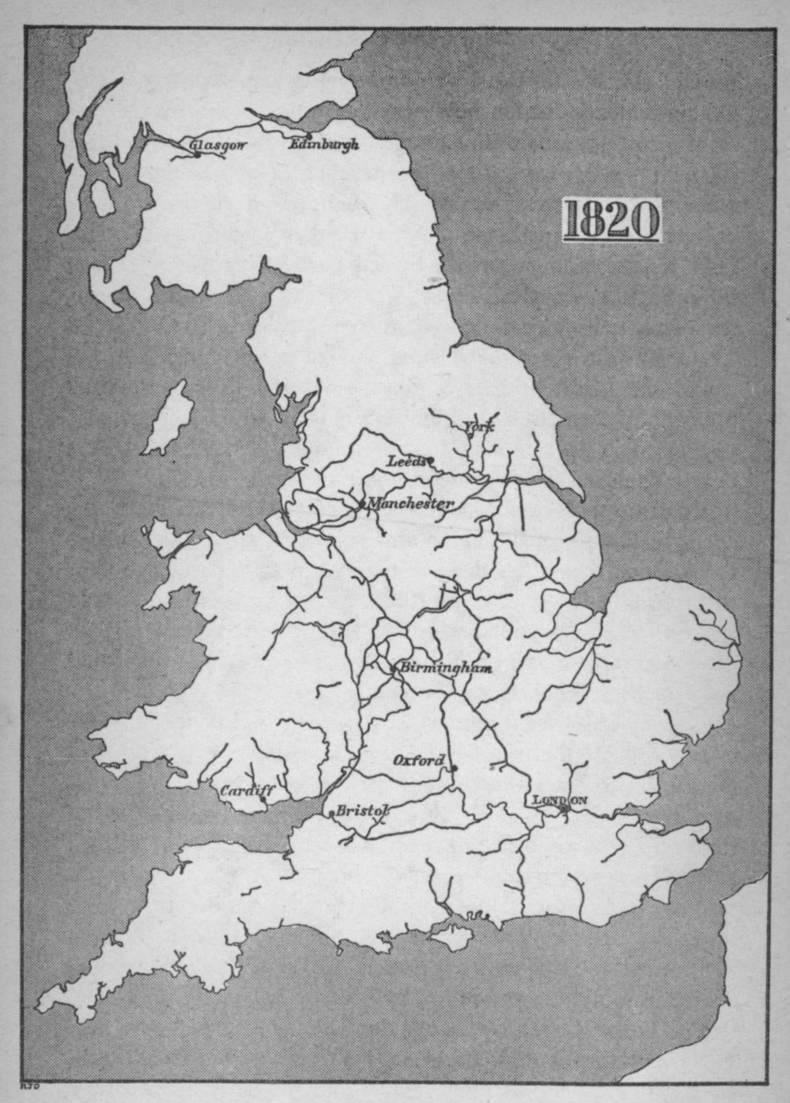

3 England, Wales and southern Scotland: inland waterways in 1820

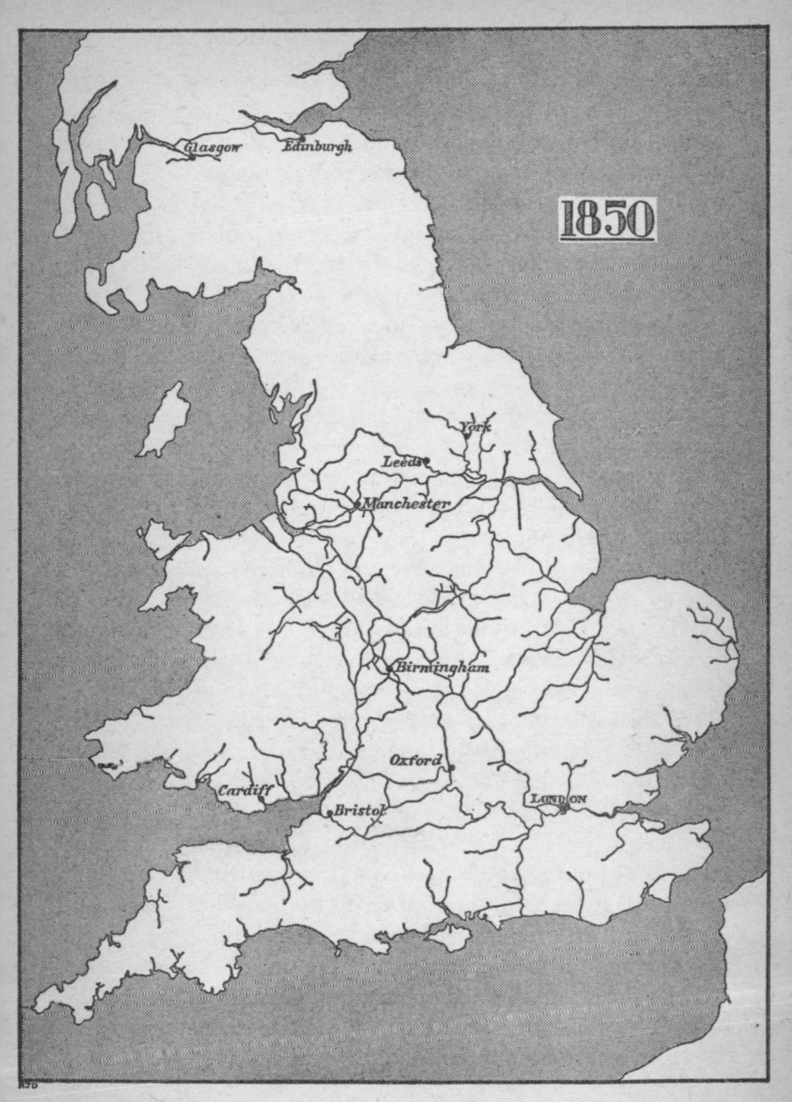

4 England, Wales and southern Scotland: inland waterways in 1850

instead one to Salford on the lines of the scheme of 1754. By the end of 1758 his ideas had crystallized into a line from Salford to Worsley and then on to the Mersey again lower down at Hollins Green on the way to Warrington. Such a canal would not only give him an outlet to Manchester for coal but also a line which would compete seriously with the Mersey & Irwell company for the Manchester–Liverpool trade. John Gilbert did the surveys, and in March 1759 he got his Act.

Before this, however, a new idea had been born. The Worsley mines had for long been drained by a sough, a culverted underground channel which carried water into the Worsley Brook. Probably it was Gilbert who thought of taking the new canal, perhaps by enlarging the sough, straight into the mines, so that the boats could be loaded as near as possible to the coal faces. And probably it was Gilbert who thought that James Brindley would be the right man to call in, for in 1753 he had built a tunnel at Clifton not far from Worsley as part of a scheme to use the water of the Irwell to power a colliery water-wheel used for pumping. Summoned, he took charge of building the canal as well as the tunnel, probably under Gilbert's supervision as the man responsible to the Duke.

Brindley was an older man of forty-three, a bachelor. Born near Buxton, he had been apprenticed to a wheelwright and millwright. He became known for mill-work, went in for engine-building and some surveying and civil engineering, but more as being the sort of man who would bring a radical mind and practical ability to a new problem. He had surveyed for a canal before, but had never built one. Work began, and by the end of 1759 about two miles had been cut from the Worsley end, as well as 150 yds of the underground canal.

Then the Duke decided on changes. He wanted his canal to end in Manchester and not Salford, because road access would be easier, and, much more important, he now wanted to drop the Hollins Green route, and instead build a canal extension down the far side of the Mersey to Runcorn, which would give him access to Liverpool for trade to and from Manchester, quite independent of the Mersey & Irwell company, and a sale for his coal in Cheshire. Therefore he wanted to change its route to take it over the Irwell to Stretford, where the line to Manchester would turn left and that to Runcorn right. A new Act was passed in 1760 to alter the canal's course to Manchester, and a third in 1762 to authorize the Runcorn line.

On March 21st, 1776, all was finished. Gilbert and Brindley had built the high embankment north of the Irwell, the three-arched masonry aqueduct over it at Barton, taken the line over Trafford Moss and made the long level canal to Runcorn, and the flight of ten locks there. Though bigger canal aqueducts had been built on the Continent, Barton staggered contemporaries: Brindley

> has erected a navigable canal in the air; for it is as high as the tops of trees. Whilst I was surveying it with a mixture of wonder and delight, four barges passed me . . . dragged by two horses, who went on the terrace of the canal, whereon, I must own, I durst hardly venture to walk, as I almost trembled to behold the large river Irwell underneath me.

At Castlefield, the Manchester terminus, the Duke built wharves, docks and warehouses. But these lay below the level of the neighbouring Manchester streets, and so, to avoid a long pull upwards for loaded coal carts, a tunnel was dug into the side of the hill at the point nearest Deansgate, into which the coal boats from the Worsley mines were run. Each was loaded with iron boxes carrying eight cwt, which were then raised to street level by a crane powered by water-wheel. Containerization has a long history.

The canal at Worsley was driven deep into the mines. From the main level, as it was called, side arms were cut, to which the coal was dragged in baskets shod with runners. Then, 56 yds below the main level, a second canal was dug, and at 83 yds below, a third. Containers of coal were winched up a vertical shaft to the main level to be put into boats there.

Then a fourth canal was made, this time higher than the main level, again linked to it by shafts, though later John Gilbert designed an underground inclined plane or railway on which boats could be exchanged between the two levels. In all, some forty-two miles of underground canal were built, portions of the system being worked until 1887.

Empty boats in trains were pulled along by men 'walking' them against the pull of a rope hooked into rings set in the roof, or 'legged' up the side cuts; full ones were propelled by opening sluices and allowing water to move them forwards on the current. Originally the tunnels, about 10 ft wide, were some 8 ft high above the water, but when in 1961 I stumbled down the ruined inclined plane to lie flat in a boat for a trip through the mine, I found that subsidence had brought the roof in places down to 4 ft. Elsewhere we were pushed through the water, yellow with iron oxide, by men lying on their backs and legging on the roof, as generations had done before them.

So the Duke's canal was built. But even before work on it had started, in Ireland cutting had begun upon the Grand Canal that was to span the island from Dublin to the Shannon. And, in England, long before his extension had reached Runcorn, James Brindley had started on the Grand Trunk (or Trent & Mersey) Canal, which was to thread a ribbon of navigable water across the country from the upper Trent to the Duke's canal, and by it to the Mersey. Nor were these alone. In counting houses and country studies, in lawyers' offices and Parliament, in the newspapers and on the bookstalls, the talk was of canals. For cheap and regular carriage of coal and raw materials meant that steam engines could be fed, factories supplied, factory-workers warmed and mines served; but more, it meant lime to improve the soil; timber, stone and slates for housing and road-making materials, and a means of moving corn and so preventing local dearths. With canal boats instead of lumbering many-horsed waggons, with steam instead of water power, what could not Britain achieve? Indeed, there would be an industrial revolution.

CHAPTER 3

They are Promoted

In the next eighty years the rivers of the British Isles were knit together, ports connected, mines served, factories provided for, by over 5,000 miles of navigable river and canal: about 900 in Ireland, the rest in Britain. The coasting trade was important; the roads carried the speeding mail-coaches, the post-chaises of the wealthy and the ponderous stages of the common man, though goods traffic was mainly local, and often directed to the nearest wharf, but the arteries that served the industrial heart of the nation were filled with water. Clear and sparkling, opaque, polluted or smelly, it foamed through the opening paddles of lock gates or over river weirs, as it carried a medley of keels, flats, trows, barges and narrow boats. It was more than water, it was money, life, progress.

And then, one day in 1830, beautiful Fanny Kemble, fresh from the balcony scene in *Romeo and Juliet* at Covent Garden, took a ride from Manchester with a North Countryman.

> 'You can't imagine it,' she wrote, 'how strange it seemed to be journeying on thus, without any visible cause of progress other than the magical machine, with its flying white breath, and rhythmical, unvarying pace . . . I stood up, and with my bonnet off drank the air before me.'

The engine was the *Northumbrian*, the companion George Stephenson, the speed 35 mph, the line the Liverpool & Manchester, and when they stopped, it was beside the viaduct where the Sankey Canal passed under the railway. Four years later, boats in the harbour of the Grand Canal at Dublin heard a puffing locomotive on the tracks of the first railway in Ireland, the Dublin & Kingstown. Water was still needed, but now mostly combined with coal to make steam.

We regard the controversy between public ownership and private enterprise as a modern one. But our ancestors were exercised whether it was right for private people to make undue profit out of providing a public utility, and their conclusion that this was justifiable has greatly influenced our canal situation today. The first true canal in England, the Exeter, had been built by the city corporation, whose successors now own its successor. Some rivers had been made navigable by private people, but there was one conspicuous case where these had failed, and public trustees had succeeded: the River Weaver in Cheshire. Parliament had decided that road improvements should only be carried out by, and the right to levy tolls granted to, public bodies of turnpike trustees. Thus, when canals were planned, it was for a time very arguable whether private companies or public bodies should build them. Should private men who had risked their own money on what was, without hindsight, a speculative proposition, benefit if it proved unexpectedly successful? And if the answer was yes, to what extent? It was to our ancestors a genuine moral dilemma.

A man like Josiah Wedgwood did not promote the Trent & Mersey Canal with the profit motive primarily before him. He did so because the object would benefit the public. Of course he would be benefited also, but probably he would not have shown such zeal and single-minded interest for his own sake alone. He wrote in terms such as, 'The public-spirited scheme', 'your Country calls', and, a year before the Act, proposed 'that the intended navigation be as free as a Turnpike road, and conducted by commissioners chosen out of the Gent$^{n.}$ in the country along which the canal shall be made'. Then, in modification, that commissioners should be 'appoint$^{d.}$ out of the gent$^{n.}$ in the Country along the Canal as a cheque [sic] over the undertakers[1] to see that the trust is executed, and

[1] Those shareholders taking the leading part.

the public not injured'. And then that the plan for commissioners 'is very strongly objected to, the objections are plausible, and abundantly verified by Experience. On the other hand, the very term Private Property is obnoxious'. It will be difficult to find people who will 'bestow so much of their time and attention Pro Bono Publico as the carrying into execution & conducting this design will require'. And so a company was chosen, though with the motto on their seal, *Pro Patriam Populumque Fluit.*[1]

In some of the early canal Acts, like that for the Staffordshire & Worcestershire, commissioners from the local men of standing, or Justices in Quarter Sessions, were given power to approve changes in tolls. An alternative was a clause in the enabling Act limiting dividends, usually to seven or eight per cent. The Trent Navigation, Derby Canal and Glamorganshire Canal proprietors were among those so limited. But, after 1800 had come and gone, such ideas of combining public control and private profit died away and directors were unrestricted. If they controlled the Birmingham Canal or the Loughborough Navigation they could pay 100 per cent; if the Trent & Mersey, 75; if the Oxford or the Coventry, 30 to 40, and so to those which paid little or nothing. The public benefit was best found, it was held, by leaving private people free to make judgements of their own interest. Good judgement would, and could, be rewarded: bad, penalized.

So it came about that our canal system was largely built by private enterprise – though with just a few exceptions. On the Continent this was usually not so, principally because improved means of transport preceded development, and were meant to encourage it, whereas in Britain they answered to a pre-existent need. In this respect Ireland followed the Continental and not the English pattern. Most waterways were built with public money, and when private companies participated, as later in the construction of the Grand Canal, public subsidies were sought and expected.

[1] It flows for country and people.

In the United States both companies and the public were concerned. The Erie, greatest of the New World canals, was built by the State of New York so profitably that, opened in 1825, in 1883 tolls upon it could be abolished, to make it a free waterway. The State of Ohio built 731 miles of canal and 91 miles of river navigation, and in Canada the military built the Rideau. On the other hand, the first American canal, the Santee, opened in 1800, and the Middlesex, from Boston to Lowell on the Merrimack, were companies; so was the great Chesapeake & Ohio. Some were private companies given state donations or special privileges; others began privately, like the Welland Canal, failed and were taken over by the State.

Then, as now, businessmen varied in type. There were some whose pockets were all important; there were some whose motives were mixed, and who combined making money with a real interest in the development of scientific discovery, either for its own sake or as a means to benefiting humanity, and in national and local achievement.

Josiah Wedgwood is an example: able businessman, he had a real love for the art that underlay his trade of a potter, and also wide scientific interests. So did Matthew Boulton, button-maker of Birmingham, who minted coins and partnered James Watt in developing the steam engine. These three were members of the Lunar Society, where, during the 1770s and 1780s, ideas were exchanged on everything under the moon, especially everything scientific. Wedgwood was the leading promoter of the Trent & Mersey Canal, Boulton a founder member of the Birmingham Canal company, and Watt, though more truly a scientist and mechanical engineer, had much practical canal experience – he had already built the Monkland Canal in Scotland, and surveyed for others.

There were also the religiously-minded industrialists, many of them Quakers: men who believed that scientific studies were also a discovery of God's handiwork, and that to contribute to industrial development was in right ordering because it would both raise the standard of living and increase opportunities for education and self-improvement by working men; industrialists who on Sundays sat down with their workmen and their families on the same benches in the same meeting houses to hear each other's ministry or to listen to their common God, and who in business meetings joined together to conduct the Society of Friends. Such men, organizers by training, rich men by ability, became expert personnel managers by the operation of their kind of religious practice.

Such were the Darbys and the Reynoldses of Coalbrookdale: 'many of the workmen in and about the works were Friends, and gathered in complete unity of spirit and equality before God, with the Darby, Reynolds, Dearman, Luccock and other Quaker families', says the historian of Coalbrookdale. 'It is not possible . . . to separate, in any of its leaders, their character and habit as Friends from their outlook and actions as employers and works managers.' These men early saw the advantages of better transport: they built horse tramroads, including the first to use iron for rails, and later a network of small canals which were lifted over the hills of east Shropshire from Shrewsbury to Donnington, Coalport, Horsehay and Coalbrookdale not by locks but by inclined planes or boat railways. They provided the initiative, much of the money and the engineering skill.

Probably private persons built the canals of Britain more efficiently than public bodies would have done, for many who promoted and later managed them had a direct interest in their success. All over the country, businessmen were avid for cheap bulk transport. The unimproved rivers were useful, but too variable to be reliable, flooding in winter and low in summer; the roads, turnpiked as many were, too expensive; horse tramroads useful, but their loads were too small in relation to their cost of construction, maintenance and haulage to make them economic except for local traffic. And so colliery owners, quarry proprietors, glassmakers, ironmasters, textile magnates, merchants, concerned themselves with canals. Lord Middleton, owner of coal mines near Nottingham, pressed the Nottingham Canal forward, as did the Earl of Moira the Ashby; the lessees of the great Caldon quarries east of the Potteries worked for a branch from the Trent & Mersey to carry their limestone: once built, it served its purpose for a century and a half. The glassmakers of Stourbridge were chief promoters of the Stourbridge Canal, and Abiathar Hawkes, glassmaker of Dudley, was for many years treasurer of that canal company, whose line extended the Stourbridge Canal upwards into the coal lands and ironworks owned by the great industrial magnate, the Earl of Dudley. In South Wales and Monmouthshire the great names in iron were canal promoters, shareholders and committeemen, almost without exception; Mackworths, Crawshays, Hills, Guests, Harfords, Homfrays. In the North were the textile men: Sir Richard Arkwright of Cromford did much to get the Cromford Canal started, and Samuel Oldknow so greatly concerned himself with the Peak Forest that the great Marple aqueduct on the line towards Manchester is often called 'Oldknow's' today. Merchants, also, especially in the North: there were twelve out of twenty-one on the first board of the Rochdale Canal. These, and hundreds more like them, were the principal driving force behind canal construction, in money, drive and administrative ability.

Associated with them were the landowners, and perhaps for the first time with any frequency landed men met manufacturing men on canal committees, and each group learned more about the other. The contribution landowners could make was enormous, for in the early days of canal promotion they and their friends controlled Parliament. And canal bills had to pass through Parliament before power to buy land compulsorily could be obtained, or to form a joint-stock company with limited liability. Parliamentary power they had to give; what had they to gain? On the whole, not personal gain, though that came in time to many of them, but the pleasurable power that comes from acknowledged leadership in an activity that is manifestly of public benefit.

Earl Gower, brother-in-law of the Duke of Bridgewater, had great estates in Staffordshire and Shropshire, and for agent Thomas Gilbert, MP, businessman and poor-law reformer, brother of John Gilbert, agent to the Duke. In 1764, with the two Gilberts, he formed Earl Gower & Company to develop coal and iron deposits on his Shropshire land, and to build a small-sized tub-boat canal taking boats about as large as those that worked within the Worsley mine.

In 1765 Wedgwood and other supporters of the Trent & Mersey Canal often visited Trentham near Stoke-on-Trent to inform and interest Lord Gower. If he would lead them, all would be well; to insist, as Wedgwood wrote to his fellow promoter, Thomas Bentley the Liverpool merchant, in 1765, 'as much as decency & propriety will permit us of L$^{d.}$ Gowers coming down into Staffordshire & PUBLICLY at a meeting of the Gentlemen in this Country . . . to put himself at the head of our design and take it under his patronage'. And, triumphantly, a week or so later, 'L$^{d.}$ Gower comes into Staffs to put himself at our head'.

He did, taking the chair at the inaugural meeting which accepted Brindley's report on the proposed canal and decided to seek an Act, and putting his name down for £2,000 worth of shares. And when the bill was referred to a committee of the Commons its chairman, fortunately, though perhaps not coincidentally, was Thomas Gilbert. It passed; the canal was built, the country near it prospered. Lord Gower never regretted his leadership, and in 1786 became Marquess of Stafford amidst the congratulations of his contemporaries.

Farther down the line of the Trent & Mersey, Thomas Anson, elder brother of the circumnavigating admiral, lived at Shugborough. He took to canals like any water-bird, subscribing £800 to the Trent & Mersey, becoming a member of its committee, supporting the building of the Birmingham Canal and taking shares in it, and generally getting involved in navigation affairs. So much did he enjoy it that he had his great house painted with a canal boat passing in the foreground.

This story could be repeated for many canals. There was the Marquess of Buckingham who sat on the Grand Junction board (his arms were incorporated in their seal) along with the Duke of Grafton, the Earls of Clarendon and Essex and Earl Spencer. There were the Earls of Stamford who had shares in the Stourbridge and Dudley canals because of their Staffordshire estates, and in the Leicester Canal by reason of interests there. There was the Earl of Powis, who greatly helped to promote the Ellesmere and the later Montgomeryshire Canal, or the Earl of Egremont, whose money so greatly supported the Wey & Arun, the Portsmouth & Arundel, and the Western Rother. The list would lengthen indefinitely, and to supplement it there would be the country gentlemen, who sometimes were also the county members of Parliament, the clergy and the doctors.

Until the 1840s, when the religious revival brought it almost to an end, the clergy's participation in canal ownership and management was notable. There was Dr Samuel Peploe, chancellor of the diocese of Chester, who helped to get the Chester Canal started, took £2,000 worth of shares, and sat on its committee for five years; or Dr George Travis, archdeacon of Chester, who gave the Rochdale company much of its early drive; or the Rev John Rocke, a committeeman of the Shrewsbury and the Shropshire canals, treasurer for the latter, and then partner in a local bank. Or the group of clergymen, which often included the vice-chancellor or college heads, who formed a majority of the highly successful Oxford Canal's committee for more than a century.

Probably the most famous doctor who concerned himself with canals was Erasmus Darwin, perhaps the original mover in the scheme that became the Trent & Mersey. He was a member of the Lunar Society, and a prolific inventor, among other things of a lift for raising canal boats vertically from one level to another; foreseeing enough to write in a poem of 1792,

> Soon shall thy arm, Unconquer'd Steam! afar
> Drag the slow barge, or drive the rapid car.

yet later remembering:

> . . . with strong arm immortal Brindley leads
> His long canals, and parts the velvet meads.

But an old friend, merely because he turns up so often, is Dr Robert Bree, senior, of Leicester, who presumably attended his patients in such spare time as he had from sitting on the promotion and management committees of the Leicester Canal – he was chairman for three years – and helping to promote the Leicestershire & Northants Union and the Grand Junction. It must have been in the blood, for his son, Dr John, was for thirty years a committeeman of the Warwick & Birmingham and the Warwick & Napton, and the Rev Thomas, maybe his brother, helped to manage the Coventry Canal.

Finally, there were the solicitors. Canal companies were accustomed to employ a solicitor as part-time clerk to deal especially with their legal business, but sometimes these men took so much interest in their canals that their efforts went far beyond their duties. John Sparrow of Newcastle-under-Lyme was one of the original promoters of the Trent & Mersey, and became its clerk. Over the next ten years or so, he was constantly at work, organizing, visiting other companies, local authorities or useful people, helping to raise money.

William Cradock of Loughborough became clerk and treasurer of the Loughborough Navigation in 1776, and of the Erewash Canal in 1777. In 1788 he passed his jobs to his son John and went on both committees until he died in 1805. Son John continued in both jobs, and on the committee of the

Loughborough for most of the time, until 1832. At his death, his son John took on for six years, died also and was followed by Thomas until 1863. The Cradocks knew so much about the two canals that they were in fact also the managers, dealing with a mass of administrative and even engineering matters in between meetings of the committee.

To the south one remembers the Bristol firm of Isaac Cooke & Son, clerks to the Bridgwater & Taunton and the Chard canals, who not only involved themselves in unwisely lending the firm's money on mortgage to the two companies but participated in boat ownership and a trade in canal-borne coal.

No analysis is exhaustive; but we may look principally to the large and lesser industrialists, merchants and landowners, and to the professional men, clergy, doctors and solicitors, for the energy to promote and the administrative ability to manage the canals in their heyday. But where did the money come from?

CHAPTER 4

Where the Money Came From

WHAT were the motives of those who invested in waterways, and who were the men? The motives were as they always have been: because investors thought canals would benefit their businesses, improve the value of land, directly or indirectly, or prove a reliable family investment; because they had a special interest in the canal as a job of work; or to make some quick money.

Take the shareholders' list of the Trent & Mersey Canal in 1766, the country's first big canal company, raising £130,000 in £200 shares. Because the canal had been promoted mainly by Josiah Wedgwood, potter, of Burslem, and Thomas Bentley, merchant, of Liverpool, businessmen were there: Josiah himself with only £1,000, his brother John with £2,000 and his cousin John Wedgwood of Smallwood with £3,000. There were some other potters (altogether investors from the Potteries subscribed £15,000), and from Birmingham Matthew Boulton, later of Boulton & Watt, with £800 and Samuel Garbett with £1,200. Among the landowners were Thomas Anson of Shugborough with £800, and that curious eccentric Sir Richard Whitworth of Batchacre with £1,000. Among investors with a special interest in the canal were the Duke of Bridgewater, with whose canal the Trent & Mersey was to connect, who put up £2,000; his agent John Gilbert with £1,000; his relation Samuel Egerton of Tatton with £3,000; his brother-in-law Lord Gower with £2,000; Miss Levison-Gower of London with £1,000; and Lord Gower's agent Thomas Gilbert, later MP and poor-law reformer, with £2,000. There was James Brindley, the engineer of the new concern, working for the Duke for small pay, but able to subscribe £2,000 to the company; and his substantial brother John Brindley of Burslem with another £2,000. Most of the other names were local, and probably serious long-term investors, like Canon Thomas Seward of Lichfield, father to Anna the poetess, the Swan of Lichfield, who invested £1,000. A few were perhaps speculators – I rather suspect Captain John Bladen Tinker of London, who subscribed £800.

If we take a more clearly industrial canal, such as the Glamorganshire from Merthyr Tydfil to Cardiff, the link between the manufacturers and their form of transport becomes clear, for the canal was promoted, largely paid for, and run by the great ironmasters to carry their iron to Cardiff. Out of £60,000 of capital subscribed in 1790, the Crawshays of Cyfarthfa and their family and connexions put up £18,100; the Harfords of Melingriffith £6,000; the owners of Penydarren £2,500, of Dowlais £1,500 and of Plymouth works £1,500. Almost half the capital therefore came directly from the five main ironworks on the line. The other subscribers were mostly local businessmen or landowners, but there was the occasional oddity – who was Samuel Lund of 86 Strand, London, one wonders? Someone who had been put on to a good thing?

The Stratford Canal shows a different picture. Promoted from that town to Birmingham in 1793 at the time of the speculative frenzy that is called the canal mania, most of its £120,000 capital was subscribed locally: £30,200 from Stratford itself, £9,500 from Birmingham, £38,600 from the rest of Warwickshire, £8,400 from Worcestershire, and £33,300 from elsewhere. If one looks at the occupations of the Stratford shareholders, however, one sees that shareholding in canals with a strong local appeal often reached an altogether lower level than, say, the Trent & Mersey, even after one has allowed for those who took shares on behalf of a wealthier speculator, or who hoped to resell at a profit before they had to meet any calls.

Apart from two widows – and one would not really think a new canal a suitable investment for them – there were four grocers, three of whom put up £1,000 each, an innholder, a yeoman, three parsons, two coopers, substantial men with £1,000 each, a plumber, two surgeons, a mealman, a perukemaker, a joiner and a mercer. Of these, one parson, the joiner and the mercer were elected to the committee.

Finally, a different picture again, twenty years later. The Grand Union Canal was promoted in 1810 to link together the Grand Junction from London towards Birmingham, and the unfinished end of the Leicestershire & Northamptonshire Union running south from Leicester. If built, it would provide through navigation from London via Leicester to the Trent. Raising money for canal shares had by now become much more organized. This particular canal, being a link in a through route and with little independent value, shows this development in an extreme case. The proposed capital was £220,000, and to begin with £70,000 worth was reserved for residents of the counties interested in its success and for landowners, the latter being given the right to subscribe for one share for every 220 yds of canal passing through their land. The promoters hoped that this chance of getting in on the ground floor would prevent opposition to the Parliamentary bill. Subscription books were opened at three banks in London, one in Leicester, and one in Market Harborough. Soon afterwards the range was widened, and additional books were opened in Birmingham, Stony Stratford, Coventry, Daventry, Buckingham, Newark, Grantham and Liverpool, brokers being paid a commission of £1 per share sold when it became fully paid. Special circulars were also sent to the shareholders of those canals with which the Grand Union was to connect. Here we have canal shares mainly sold for investment purposes to people unconnected with the concern or those who ran it, though with a large minority of shareholders who were local or concerned with neighbouring canals.

A very few canals were built for the capital authorized in the original Act. Nearly all required more money. With some, of course, the prospects were so good or the backers so rich that it was easy to raise more funds. With others it was very different. A half-built canal was worth nothing in itself, and was therefore difficult to mortgage unless so near completion that the sum lent would clearly finish it. Therefore money had to be got by calls upon the original shareholders beyond the nominal value of their shares, or by issuing new shares at attractive rates to old or new holders.

The Stratford-upon-Avon Canal was typical. The authorized capital, as we have seen, was £120,000, with power to raise £60,000 more. The Act had been passed in 1793, and by 1802 £150 had been called on the £100 shares, and authority was exhausted. Only half the canal had been built, though fortunately the more useful half, which became part of a through route from Dudley towards London. Nothing further could be done until 1809, when under a new Act £60,000 was raised in £30 shares, ranking *pari passu* with those on which £150 had been called, and a further £30,000 of them in 1814. By 1815 the company was forced to raise £35,000 in annuities at rates up to 10 per cent. In June 1816, having incurred £20,000 worth of unsecured debts as well, the work was finished.

Preference shares were seldom offered by canal companies, but one often finds the loan note convertible to shares within a certain period. Prior charges, whether debentures, mortgages of property or tolls, loan notes or bonds, were almost always raised from private individuals, usually shareholders, hardly ever from institutions, except occasionally from insurance companies at the very end of the canal age. Most such charges were safe enough. A few companies could not pay the interest – like the Newcastle-under-Lyme, or the Thames & Severn, which persuaded their creditors to take shares for most of their debt and interest arrears – or could not repay their debts on the due dates, like the Worcester & Birmingham, which thereupon had a receiver appointed.

Another source of canal finance was public money. This began in Scotland where, under an Act of 1751–2, the Forfeited Estates Fund was created by the transfer of property whose owners had been attainted for participation in the rebellions of 1715 and 1745. Its produce was to be used for 'civilizing and improving the Highlands of Scotland and securing the peace and loyalty of its inhabitants'. In 1784, when the partly-built Forth & Clyde Canal across the waist of Scotland had come to a stop for lack of money, the Government authorized a loan of £50,000 from the fund to finish it. Later, they lent what the Forth & Clyde repaid to the Crinan Canal Company, so that their waterway in turn could be opened.

In England, the Poor Employment Act of 1817 was passed to enable Exchequer Bill Loan Commissioners to lend money for works that would employ the poor. Two important canals, the Regent's and the Gloucester & Berkeley (the present Gloucester & Sharpness Ship Canal) would probably never have been finished without such help, which was also given to a number of other waterway concerns in Britain and Ireland. Between 1817 and 1828, the Commissioners lent £623,000 for navigation improvements.

Two canals were actually State-built for national defence reasons, the Caledonian in the north of Scotland and the Royal Military in Kent, and in 1848 the State also took over the Crinan, which it had managed for some years before that. Others were built by local authorities, like the oldest of all, the Exeter Canal. Another was Beverley Beck in Yorkshire, improved by Beverley Corporation under an Act of 1726. In the early days of the canal age, corporations often subscribed to canal enterprises and took a seat on their boards. But municipal participation had ended by about 1800, and did not revive until Manchester Corporation stepped in to enable the Manchester Ship Canal to be completed, and in the first decade of the present century Nottingham Corporation began to improve the Trent Navigation.

Banks were an important source of finance in the days before limited liability, but here it is difficult to distinguish between the banker as an individual shareholder and the bank as an institution, for even when a bank was appointed treasurer of a company, it was always in the names of certain individual partners, usually shareholders also, who had to give personal security for the safety of the company's money. Such individuals often had other occupations also; industrialists, perhaps, like the Reynoldses of Ketley.

The first treasurers of the Worcester & Birmingham Canal do not seem to have been bankers when they took the job, though they are described as such soon afterwards: Isaac Pratt, probably a coal merchant, and Thomas Hooper, whom one thinks to have been a rich, learned and not very financially-minded man in private life. Pratt held 50 shares of £100 and Hooper 48, and each had to give a bond for £10,000 and find a surety for £5,000. Pratt resigned in 1796, and two years later it was found that Hooper, who had taken on alone, owed the company some £14,000. One of his sureties paid £5,000, and most of the rest was recovered in time.

Hooper was succeeded by a banker, Isaac Spooner, holding five shares. As a banker he seems to have been a disappointment to the company, who were chronically in need of money to finish their canal (it took until 1815), and in 1803 he was replaced without notice by the Birmingham bank of Wilkinson, Startin & Smith. From 1811 onwards, with Isaac Spooner now as chairman, a drive was made to finish the canal. Money was raised by shares and debentures, and in the next eighteen months the bankers advanced £27,000. Then, though publicly thanked, they refused to lend any more, and the company had to fall back on raising annuities and running up debt. From the annuities they repaid their bankers, who had gone so far as to get a clause in the company's Act of 1815 enabling them to demand the compulsory sale of land bought for reservoirs unless the loan were repaid by the autumn – which it was.

EARLY DAYS IN EUROPE: (*above*) Erlangen lock on the Ludwigskanal; (*below*) the flight of seven staircase locks at Béziers on the Canal du Midi

EARLY DAYS IN EUROPE: the Canal de l'Ourcq in Paris, 1830

Ferdinand de Lesseps

TWO AQUEDUCTS: (*above*) Brindley's of 1761 that carried the Bridgewater Canal over the Irwell, just before it was demolished to make way for the Manchester Ship Canal; (*below*) Pontcysyllte, carrying the Ellesmere Canal 121 ft above the Dee, was the creation of the company's engineers, Jessop and Telford. Opened in 1805, 1,007 ft long, it still carries pleasure craft to Llangollen

MAINTENANCE PROBLEMS: (*above*) a cross-dam in position in Blisworth tunnel; (*below*) a spectacular canal burst at Prestolee on the Manchester, Bolton & Bury Canal. Note the containers in the boat

CONTRASTS: (*above*) an Upper Rhine barge of 1531; (*below*) an industrial scene in the canal age

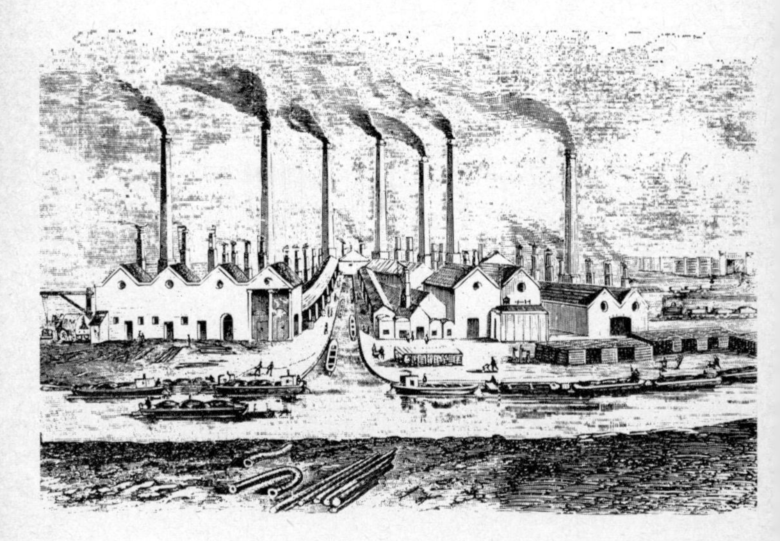

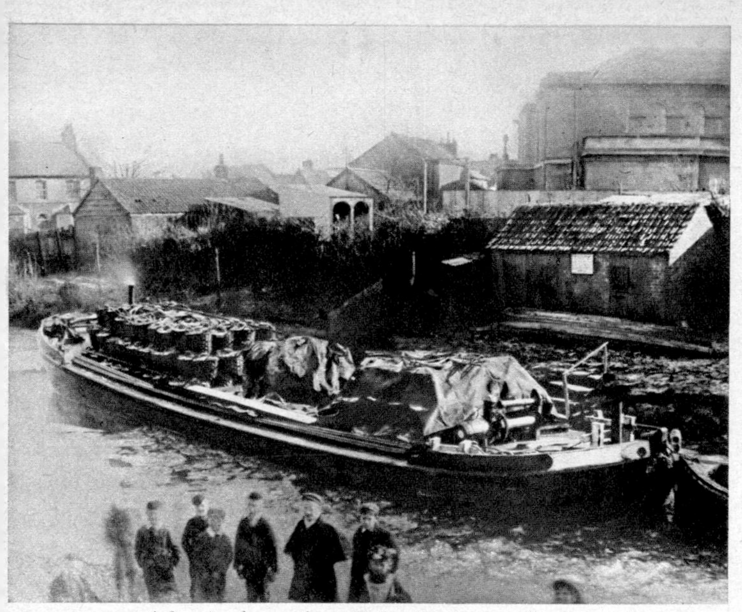

CARRYING: (*above*) a barge from Bristol arrives at Devizes wharf on the Kennet & Avon Canal; (*below*) loading limestone brought down the tramroad from Caldon Low quarries at Froghall on the Trent & Mersey Canal

CANAL-CONNECTED PORTS: (*above*) Ellesmere Port *c* 1907; (*below*) the basin and entrance lock of the Bude Canal in Cornwall in the 1890s

Because of their unhelpfulness, the company's bankers were now replaced by Smith, Gray, Cooper & Co of Birmingham. This bank, later Gibbins, Smith & Goode, failed in 1825, perhaps through being too helpful too often, and were succeeded by Spooner, Attwood & Co, the Birmingham banking company. Isaac Spooner being still chairman, and Richard Spooner a member of the committee, once again the company's banking arrangements had returned to prominent canal directors.

The history of this canal illustrates a point perhaps worth making. Early canals, like the Trent & Mersey, often took a long time to build for engineering reasons – in that case eleven years. But some later waterways, substantial in size and cut during the Napoleonic Wars, took even longer for financial reasons. Because of rising prices and inflation they cost more than had been estimated; calls were often in arrear because shareholders just had not got the money, and if pressed too hard would rather forfeit their shares; additional money was difficult to raise, and then only at a rate which severely penalized the original shareholders, and extensive bank help was usually out of the question because money would have been tied up for far too long: in 1803, for instance, when the Rochdale Canal was nearly finished, the bank refused to advance more money, and demanded a reduction of their overdraft; the harassed committee then minuted that 'the concern is going to ruin for want of the arrears being paid up'. To find money, they had to issue low value shares ranking *pari passu* with the much higher original ones. Therefore work went on only to the extent that money could be raised each year. The Worcester & Birmingham took from 1793 to 1815. Other canals similarly delayed were the Gloucester & Berkeley Ship Canal, from 1793 to 1827; the Huddersfield Canal, 1794 to 1811; and the Kennet & Avon, 1795 to 1810.

At a time of difficulty in raising money, a canal company might offer its account to a bank that would lend some. In 1803, after it was completed, the Swansea Canal Company asked their bank for a loan of £3,000, and were refused. They then offered the alternative of lending it for five years at five per cent interest on mortgage of the tolls, or losing the account. The bank lent the money. In 1802 the Monmouthshire Canal Company tried to borrow £10,000 from Edwards, Templer & Co of London as a condition of offering them the treasurership. The firm refused, and Buckle, Williams & Co of Chepstow then agreed on the same condition. An odd result of competition for the business of a profitable canal was the rule made by the Neath Canal Company in 1822 that they would bank in alternate years with Williams & Howland of Neath; and Haynes, Day & Haynes of Swansea.

The Brecon Old Bank was one of the best known of early country banks. The four partners from 1796 onwards took £7,600 worth of shares in four Welsh canals, seemingly as part of a deliberate policy of investing in Welsh industry. In the mid-1820s, these were worth about £13,000. The bank's partners held the treasurerships of the Brecknock & Abergavenny Canal from 1793–7 and 1812–28, the Glamorganshire from 1793 onwards, and the Hay Railway, a horse-tramroad branch of a canal, from 1812, and were during the period on the committees of four Welsh canals and the tramroad.

Let us now look at an important line of canal which owed much to the energy and financial help of a banker and his associates. William Praed, senior, a member of the banking family of Praed & Co of Truro and an MP, living in 1792 at Tyringham near Newport Pagnell, was one of the principal promoters of the Grand Junction from London to Braunston to join the Oxford, Coventry and other canals to Birmingham and beyond. He became chairman when the company got its Act in 1793, a post he held till his death in 1823, and invested in 55 £100 shares, which by 1804 had increased to 194. His son William Tyringham Praed in 1801 formed a banking partnership with Philip Box of Buckingham, who had been the canal's treasurer since 1793, and who had increased his own holding of shares from 22 to 51½. The new firm of Praed & Box at once became treasurers. Soon afterwards Praed & Co opened in Fleet Street as London bankers, and in 1809 young William joined his father on the canal committee, where he stayed until his death in 1846. Young William pushed the canal's interests towards Leicester by promoting the Grand Union to connect it to the canals leading through that town to the Trent. William sat on the Grand Union committee from its formation in 1810. Praed & Co were of course treasurers, and were concerned in promotion with the banks of Pares & Heygate of Leicester and Inkersole & Co of Market Harborough. Like William Praed, James Heygate of the former and Thomas Inkersole of the latter were on the promoting committee.

To summarize a little about one's impression of bankers. They did not make a notable contribution to canal financing. But canals were of great use to them, by providing opportunities for personal investments which sometimes lost money but more often made it, and which were less risky, and easier to raise money upon, than interests in industry. Grand Junction shares to the Praeds were as solid as the best blue chips are to us. Again, bankers often took a prominent part in promoting canals, partly, of course, from a natural wish to increase the prosperity of their neighbourhood or because of preceding links with local industry or commerce. In return, they were sometimes rewarded by being made treasurers, with money to be earned from keeping the company's balances, making loans, selling shares, and paying dividends. But one suspects that they also found the position of canal treasurer, or failing that of committee member, useful in the business of banking, for it must have given them much inside information and many useful business contacts. After all, the products of almost every major firm in England were carried on the canals, and their business was written over every page of committee and account books, or could be discovered from inquiries among the carriers.

So, from businessmen and landowners, doctors, widows,

tradesmen and parsons, and from banks, came some £20 millions to build 2,600 miles of canal in England and Wales between 1760 and 1850. Some of the waterways they financed are still useful today, and all are monuments to the energy and industry of those who promoted and paid for them. Monuments, also, to the men who built them.

CHAPTER 5

Engineering and Construction

CANAL engineers owed much to the earlier drainage and river men, and to the accumulated experience of building mill weirs, streams, ponds, races and tumbling bays. Those in Britain, however, seem to have learned nothing from previous Continental experience, but to have taught themselves again what men like Riquet, builder of the Canal du Midi, had done a hundred years before.

The great name is Brindley, who only in fact built part of two canals, the Bridgewater and the Trent & Mersey – both were unfinished at his death – and then became a quickly-moving, temperamental, highly-paid consultant: 'my master Brindley never paid half the attention to all the canals he was concerned in as I pay to this single one; he neither set out the work or measured it as I measure it, yet he received near £2,000 a year', as one of his former assistants, seeking more pay, wrote rather sadly.

Brindley trained his assistants well, by the simple method of letting them do the work, with only the briefest of visits from himself. He and they, but mostly they, built many of the early canals. Hugh Henshall, whose daughter he married, the experienced man who finished the Trent & Mersey and Chesterfield canals for him; Robert Whitworth, who helped to join Forth to Clyde, Mersey to Aire over the Pennines, Thames to Severn; Thomas Dadford, who with his sons built the Staffordshire & Worcestershire and its connexions, and most of the South Wales canals; and Samuel Simcock of the Birmingham and the Oxford. Outside Brindley's galaxy we can glance at Thomas Telford, giant of the Caledonian and the Shropshire

Union main line, road and bridge builder also; John Rennie of the lovely aqueducts; Benjamin Outram of canals in Pennines and Peak, and horse tramroads everywhere; Josiah Clowes, self-taught tunnelling specialist; and James Green, builder of inclined planes and canal lifts on the small and hilly waterways of the South-west.

1 Signature of William Jessop, at the end of a manuscript report on the River Trent Navigation

Greatest of all, I think, but less famous because more unassuming, was William Jessop, pupil of John Smeaton, also a canal builder, who flung the Grand Junction half across England, the Ellesmere into Wales, and the Grand Canal from Dublin to the Shannon. Jessop, universally sought after, never quarrelled with, never valuing himself at his true worth, accepting blame unquestioningly for a mistake: 'I have been examining the disastrous state of the Derwent Aqueduct . . . the failure has happened for want of a sufficient strength in the front walls and I blame no one but myself for the consequence.' He goes on to offer to pay for the rebuilding, for 'painful as it is to me to lose the good opinion of my Friends, I would rather receive their censure for the faults of my head than my heart'.

Theirs was a rushing life on horseback, or in coach, working in inn parlours, riding through the rain ('there has been very little weather yet in this country fit to stand out to level and survey, though I have done it more perhaps than was prudent, being very often much wet'); sometimes ill ('I am still obliged to the continual use of Fermentation and there is formed a lodgement of matter on the Cheek Bone which is out of the reach of the surgeon . . . I am still confined, but hope next week to be able to get out with my Head wrapped up');

struggling to supervise resident engineers and contractors strung out on a long line of cutting ('it is the closest business I ever had, to keep such a number of workmen in regular order, as they are employed on such a variety of works that are now upon hand and lie so wide'), always called to be there when just arrived here. The life killed Brindley when still only 55: 'I think Mr Brindley – the *Great*, the *fortunate*, *money-geting* Brindley, an object of Pity! . . . He may get a few thousands, but what does he give in exchange? His *Health*, and I fear his *Life* too.' It compelled Smeaton to retire, 'seeing no prospect of materially bettering my condition, but the certain prospect of destruction to my constitution viz, subjecting myself to the same hurries and fatigues that had occasioned the attack'. But it got the canals built.

Most British canals were cut before the maps of the Ordnance Survey had appeared. Local maps existed, but normally engineers and their surveyors had to work out alternative routes, doing their own levelling, until they found a satisfactory line. This, with estimates of cost, traffic and revenue, had to be put before a promotion meeting. The businessmen of the committee were capable of making the estimates of traffic and revenue, and of briefing the engineer upon the route the canal should take to obtain the most traffic. But he had to choose the exact line it should follow.

In the days when the early canals were built, savings available for long-term investment were scarce. Construction costs had to be kept to a minimum, and therefore expensive engineering works were avoided. A canal wound along a contour until it was necessary to insert one or more locks to raise or drop it to another contour, to wind onwards once more. Walk along the Trent & Mersey, and observe how seldom you will find even a small embankment or cutting. The extra length cost far less than the works that would have been necessary to avoid it, and sometimes was a positive advantage, for more collieries or other sources of traffic could be served. Once built, the advantage of water transport over road carriage more than offset the extra costs of men and horses on the more winding line. But later, turnpike roads improved, the first railways appeared, speed and lower carrying costs became more important. And so canals were shortened, notably the Oxford and the main line through Birmingham, and new ones were built, like the Macclesfield and the Shropshire Union main line,[1] which ran as straight as possible, with great embankments across valleys and cuttings through hills. But financially they were not successful: the extra construction costs meant tolls too high to be competitive.

Besides cost, the engineer will have water supply in the forefront of his mind. What stream can be led to his waterway without affecting the supply to mills or gentlemen's parks? What sources are there for the summit or topmost level? And if reservoirs will be needed, where can they be put? For canals are expensive in water. Take a canal over a watershed, like the Leeds & Liverpool or the Rochdale over the Pennines, both with broad locks, using 50,000 to 60,000 gallons each time one was worked.

Imagine two barges following each other from east to west through the last lock up to the summit. Boat I finds the lock full, because another boat ahead is going the same way. It has to be emptied, taking 50,000 gallons off the summit level; then our boat enters, using the same quantity again. Boat II has to empty and then refill, using another 100,000 gallons. At the other end of the summit Boat I finds the first lock down to be empty, as the boat in front has left it, and therefore fills it, goes down, and leaves it empty. Boat II has to repeat the process. Two boats passing have therefore taken at least 300,000 gallons of water off the summit. No wonder such a canal needed enormous reservoirs, though of course water could be saved, for instance by 'working turns', that is, making craft go past locks alternately up and down, or by pumping water up the locks again. The Leeds & Liverpool had seven reservoirs, with a total capacity of 1,173 million gallons; the

[1] The Birmingham & Liverpool Junction as it was then.

Rochdale seven also, holding 1,529 millions. Water was valuable in Britain when the canals were built, owing to its extensive use for industrial power. Because of this, and the scarcity of capital, waterways had to be constructed as small as the estimated traffic would warrant. Hence our legacy of undersized canals, too limited in capacity for modern needs, and yet not all of the same dimensions.

A lock in Britain usually raises or lowers a boat between 6 and 10 ft – the biggest is about 16 ft, the minimum a few inches. Its placing was generally dictated to the engineer by the natural features of the line. If this fell steeply, he would need a flight of more than one. If very steeply, he would have to build them as a staircase, with the bottom gate of one lock acting also as the top gate of the next. But sometimes their siting was determined for special reasons – where water supplies could be led into the canal, how intensive short-haul traffic could best be given a run lock-free, or with the minimum number; and how, especially in towns, they could best be supervised. Single locks in succession were usually spaced far enough apart for the level of each pound not to be seriously affected by their working, and built with about the same rise and therefore capacity. Thus no pound would be drawn down because the lower lock used more water than the one above, or overflow because it used less.

The engineer, in laying out his line, had to take account of the likely cost of land, and be careful as far as possible to avoid houses or valuable buildings. Many a canal Act specified just how close a canal might pass to a named house: 'nothing in this Act . . . shall . . . empower the Company . . . to . . . approach . . . the said line of Canal nearer than . . . One Hundred Yards . . . from the Parsonage House and Garden Wall of . . . the said Rectory of Tidcombe', or which side of the canal the towpath should be along certain stretches, or that mooring should be prohibited near gentlemen's houses and parks. Compulsory power of land purchase might not include houses: the Grand Western company, in the same Act as I have quoted, was prohibited from taking any house or building, garden, orchard, yard, park, planted walk, avenue, lawn or pleasure ground, without the consent of the owner or occupier, except in specified cases listed in a schedule. For though a company's Act gave it compulsory purchase powers within the authorized limits of deviation laid down in its Acts, the price to be paid was problematical. If buyer and seller could not agree, they could agree on an arbitrator. In the last resort they could go to a quorum of the Commissioners named in their Act, who would be composed mostly of landowners, unlikely to set a price too low for fear of creating a precedent, and finally, to a jury summoned by the sheriff, whose award was binding.

If the engineer had done his work well, the promotion meeting accepted his report and plan, the bill passed, and he was instructed to start work, money being provided by periodic calls on shares, upon which the company usually paid fixed interest until the job was done. And here canal companies, like many other concerns later, found how difficult it was to control their professional and his subordinates. Those did best who had, perhaps, a chairman, perhaps a small group of committeemen, willing to give time to being available for consultation, and about on the line with their eyes open. Those did worst who left the engineer unsupervised, for even the best of men does better in the presence of those to whom he is responsible.

The early canals were not usually built by a single contractor, for contracting firms of sufficient size did not then exist. Instead, the engineer would let a number of contracts to separate small contractors or hag-masters, for cutting sections of canal each a few miles long, at so much a cubic yard of digging,[1] plus puddling and lining at so much a cubic yard, and others for work such as locks, bridges, tunnels, and the necessary

[1] Before the French war inflation, the price was about 3*d* a cubic yard. It then rose slowly to about 6*d* at the end of the war, and ten years later was down to 4½*d*. Deep cutting, or removing the soil more than about 20 yds, was extra.

buildings. Quite a number of canals were built by direct labour employed by the company and superintended by the engineer, and some by professional canal engineers acting as contractors – the Glamorganshire from Merthyr Tydfil to Cardiff was so constructed by the Thomas Dadfords, father and son, and Thomas Sheasby. Later, civil engineering contracting firms appeared, such as Joliffe & Banks, who built the Knottingley & Goole Canal of the Aire & Calder in the 1820s and the Ancholme Navigation in the 1830s.

When the earliest canals were started, men were still bound to their work. A contractor on the Staffs & Worcs Canal advertised in May 1767 in the Birmingham newspaper a list of men, with their heights, who had absconded, and threatened legal action against anyone employing them. The engineer of the Chesterfield Canal in 1773 was empowered to deduct from a man's wages the expenses of bringing him back if he had run away. By the 1780s, however, such binding was no longer feasible, though efforts were made to prevent labour being poached. The Neath Canal in 1791 announced that they would not hire anyone from works in Glamorganshire without a discharge from his manager, because men were moving about in hopes of better wages 'which has occasioned much Inconvenience and Loss to persons engaged in such Concerns'.

During the French Revolutionary Wars, with their steeply rising food prices, canal digging often stopped during harvest. In 1793 the Ashby Canal's engineer was told to reduce labour during these months 'in order to let them go off to Harvest Work for the benefit of the country', and in 1794 cutting of the Peak Forest was to begin 'as soon as the Corn Harvest shall be got in'. Previously it had been unnecessary to advertise for labour, but now minute books order 'that an advertisement be inserted in the Leicester and Northampton Newspapers for Bricklayers, Brickmakers and Diggers'.

Before these wars, there was not yet sufficient public works contracting in existence for a class of professional contractors' men to have been created. Therefore it seems probable that the older canals were cut by men who worked in the neighbourhood of their homes, having been recruited by small local contractors known to them. But by the later canal period, a class of professionals had begun to work on canals and other public works. These professional navigation cutters or navigators (navvy is a later word) would then be found supplementing the local men. Where, however, there was a single contractor for the whole work, or when it was being done by direct labour, or in a tunnelling contract, it is likely that most of those employed would be professionals. But we still know little about the men who built our canals – far less than has been set out for us in Mr Terry Coleman's book, *The Railway Navvies*, for a later time.

The engineer arranged continuous inspection by 'overlookers' of the quality of the work done and its measurement for progress payments to the contractors. On direct labour jobs, the men were paid monthly or fortnightly, but could draw weekly advances or 'subsist' payments. Sometimes bonuses were also paid for special efforts. The men cutting the Oxford Canal in 1770 seem to have been roughish types, for the committee

> ordered that 2 Pairs of Pistols be purchased . . . to be delivered into the hands of the paymaster to this Company for the use of the Company's Agents.

Accidents were bound to occur. As the men were usually employed by small contractors, the company was not responsible, but some helped to form sick clubs, financed by weekly contributions from the men. The company might add donations to the sick club, to the local hospitals to which accident cases were taken, or, if there were no sick club, direct to the men or their widows:

> James Whitworth, the person who has Hurt his Hand at the Works, be allowed 5*s* per Week . . . till he becomes able to work.

or

> a Guinea to be given . . . to the Widow of the person killed in the Works . . . towards the expenses of his Funeral.

Under the principal engineer, who was probably not full-time, there was a resident, and on big canals like the Grand Junction, up to three sub-engineers in charge of lengths of work, and under them master masons and smiths. Given that the canal had to be provided on one side with a towpath, the engineer started by pegging out his line, having in mind the object of moving the least amount of soil the shortest distance, and so balancing cutting and embankment that, when the job was done, no pits whence soil had been dug, or spoil banks of excess material, should remain.

Top soil had first to be removed for later return to the completed banks and neighbouring land. Digging was done by hand, using wheelbarrows on planks, and temporary horse tramroads for moving soil longer distances. In a deep cutting, where a man and his barrow had to be got up the sides, rings on the end of ropes could be slipped over the barrow's handles, and it could then be helped up by a horse at the top walking away. Should cutting be through clay or other watertight soil, no lining would be necessary; otherwise the canal bed must be lined or puddled. Puddle, said an instructional account written in 1805

> is a mass of earth reduced to a semifluid state by working and chopping it about with a spade, while water just in the proper quantity is applied, until the mass is rendered homogeneous, and so much condensed, that water cannot afterwards pass through it, or but very slowly. The best puddling stuff is rather a lightish loam, with a mixture of coarse sand or fine gravel in it; very strong clay is unfit for it, on account of the great quantity of water which it will hold, and its disposition to shrink and crack as this escapes.

This puddle was then spread in layers over the bottom and sides of the canal excavation to a thickness of 18 ins to 3ft depending on the porosity of the soil, and covered with a layer of ordinary soil or second-quality clay about 18 ins thick. The canal then had to be filled with water before hot weather could damage the lining. Filling with water also enabled working boats to be used to move soil and construction material from place to place.

Meanwhile brickworks had been set up locally to make bricks from local clay, or, in stone country, orders given to local quarries. Timber had been bought for lock gates and beams, and for building construction. Lock pits had been dug, and bricklayers, masons and carpenters would be at work on the lock chambers, on aqueducts or culverts to take the canal over streams, and bridges to carry roads or reconnect portions of farms divided by the cutting. Such accommodation bridges could be permanent brick or stone structures, but more often timber swing or lifting bridges in various designs. If a river were being made suitable for barges, new bridges would probably be fixed structures. But those already in position were a problem. The contract of 1791 for making the Soar navigable from Loughborough to Leicester stated:

> The new bridges to be nineteen in number, three of them to be twenty feet wide in roadway, seven of them to be fifteen feet wide, and nine of them to be twelve feet wide; the ascents to the bridges not to be steeper than three inches in a yard. The old bridges that are to be altered by putting two or three arches into one, must have an opening for the passage of boats not less than sixteen feet in width, and the underside of the crown of the arch to be not less than ten feet and six inches above the surface of the water in a full Pond.

Small aqueducts to carry the canal over streams or by-roads would be masonry or brick structures big enough to carry the canal in a watertight lined channel, and its towpath beside it, though the channel might be narrowed to save expense. Bigger ones in masonry, such as the famous examples that carry the Lancaster Canal over the Lune near Lancaster, the Peak Forest over the Goyt at Marple, or the Kennet & Avon over the Avon at Limpley Stoke near Bath, masterpieces of design by their engineers, were built on the same principle. In the 1790s, however, iron aqueducts were first introduced by Benjamin Outram on the Derby Canal at the Holmes, in the city of Derby, and Telford at Longdon-on-Tern on the Shrewsbury Canal, and later by Jessop and Telford on the famous Pontcysyllte aqueduct of the Ellesmere (now usually called the Welsh) Canal near Llangollen. Others, such as Chirk on the Ellesmere Canal and those on the Edinburgh & Glasgow Union Canal, were hybrids, enclosing an iron trough in a framework of masonry.

Tunnelling wrote indelible lines of worry on many an engineer's face. Harecastle on the Trent & Mersey, the first long one to be built in Britain, 2,880 yds long, took over eight years to cut; Morwelldown tunnel on the Tavistock Canal in Devon, 2,560 yds, took thirteen years; and Standedge, 5,456 yds, the longest to be built in England, over sixteen years. In England and Wales about 42 miles of canal tunnel were built; in Scotland only one short one, in Ireland none, in the United States a few only, and short. Europe, especially France, has a number, including the 7,301-yd long Riqueval tunnel on the St Quentin Canal, and the longest on the world's waterways, the 7,789-yd and very large Rove tunnel on the Marseilles-Rhône Canal, over four miles long. But England and Wales probably built a greater length of canal tunnel than the rest of the world together.

Before canal tunnels were needed, the art had been learned in mining and mine drainage. With only that geological knowledge which a few trial borings could give them, engineers worked largely on hit or miss, and sometimes they missed. The first task was to stake out over the hill an exactly straight line, to mark the course of the tunnel. Then, because tunnelling would almost certainly reveal springs of water, a stream had to be found below the proposed line, and a channel made from it to a point below the mouth of the tunnel. Then a start would be made on driving a separate small drainage heading beneath the actual line. Should water be found during excavation, a side-heading would be driven towards its source to carry the water into the heading below the tunnel works, and so away to the stream.

Simultaneously with driving the heading, shafts about 8 ft in diameter and 150 yds apart would be dug from the surface of the hill above to intersect the centre of the main tunnel line. These shafts were often dug after the manner of wells, by steining the shaft – that is, building a few feet of brickwork upon a curb, and then excavating from underneath the curb and letting the steining sink, building fresh brickwork on it as it did so. If strong springs were struck, steam pumping engines might have to be used, as at Standedge. Excavated material could be drawn up the shafts by horse gins, and bricks for lining, wooden centres, and other things needed by the tunnellers, sent down. Such a shaft also provided ventilation to the workings, sails being erected at the top to direct the wind downwards, and the shaft being divided to provide both up and down draughts. Updraught could be helped by a fire lit underneath one half of the shaft. Later, after two shafts had been connected underground, one could be used for up-draught and the other for down. Nineteen such shafts were built for Blisworth tunnel on the Grand Junction Canal, twenty-five for Sapperton on the Thames & Severn.

Given an exact line over the top of the hill, this line could be accurately transferred to a string across the top of a shaft. Then from each end of the string, lines could be dropped to the tunnel level, and steadied by plumb-bobs hung in water. The two lower ends of the plumb-lines would then be connected with string to give the exact underground direction needed. At the same time the engineer, working from the levels he had taken over the hill, would transfer the exact depth of the invert or bottom of the tunnel to a mark on the brickwork of the steining. So, with the direction and level fixed, digging or blasting with gunpowder started outwards from each side of the shaft, the men working by candlelight. As each yard was dug, timber ribs for centrings would be assembled and bolted together. Laths were then fixed outside them and round these the brickwork would be built, the space between the excavation and the brickwork being then filled with rammed clay. Where a tunnel was driven through rock, parts were sometimes left unlined as in Standedge.

To take Blisworth as an average tunnel: when built the internal width was 16½ ft, the height from the crown of the arch to the bottom of the invert 18 ft, 7 ft of which were below water line. The side walls were segments of a circle of 20 ft radius, the top arch of one of 8 ft; at the bottom there was a curved invert. The top and side walls were two bricks thick, the invert 1½ bricks thick. As the years went by, most tunnels were affected by minor earth movements, some by mining subsidence, or the action of springs behind the lining. The walls and invert became distorted, or the roof became lower. Often one can trace the effects of movement in the brick courses of the lining, though rebuilding will have taken place after serious movements. The narrow Lappal tunnel on the old Dudley Canal, much of it through soft rock, was continually in need of maintenance; the old Harecastle tunnel, Butterley on the Cromford Canal and Norwood on the Chesterfield Canal were closed after mining subsidence, and the new Harecastle tunnel built by Telford and opened in 1827, also suffers from it, as those discover who take a pleasure cruiser through.

Locks vary much in size, but nearly all are of the same pattern, parallel brick, masonry or timber walls being enclosed within gates at either end. These used very occasionally to be hung from pintles on the gate-posts like a farm-gate; if so, as on the River Lark or the Essex Stour, a high galley beam was installed to connect the post tops above barge level to prevent them being pulled inwards by the weight of the gates.

Gates are usually built with a rounded heel-post, the bottom of which pivots in a recessed metal pot let into the sill, against which the feet of the gates rest. The heel-post moves in a semi-circular quoin cut in the sides of the lock, and is held at the top by an iron strap. Beams run along the top of each gate and are continued over the lock side, both to balance the weight, and to provide a means of moving the gate easily.

Most locks have a pair of gates, meeting at an angle, the 'V' of their shape pointing towards the higher water level so that its pressure will keep them shut. Narrow locks, however, may have a single gate which turns to fit into a slot in the opposite wall of the lock and so lie straight across it. Small lock gates are usually wooden, though a few are of iron or steel. On bigger canals gates are usually of steel.

In most gates there are openings below water level, which can be closed by sliding a wooden paddle across them, this being worked by gearing consisting of a vertical rack engaging a pinion (fitted with a catch) operated by a hand windlass. By using these paddles, and also others working in openings made in the structure of the lock behind the gates, locks can be filled and emptied. This basic pattern was sometimes varied. Locks could be diamond-shaped, to enable extra water to be passed through the lock, or because engineers thought them stronger that way. Old ones on rivers might have sloping turf banks instead of brick or masonry walls, barges being held within a wooden framework to prevent them going aground. Instead of 'V' or mitre-gates, some English locks had – and have – vertically or radially rising gates, as on the Shrewsbury Canal or the River Nene, while in the United States the drop type was sometimes used, which fell flat on to the canal bed. Nowadays many other designs of gate are used around the world, while lock structures often admit the water all down the sides (as do the locks on the Warwick line of canal) or even from beneath the floor. On bigger waterways, in Britain and

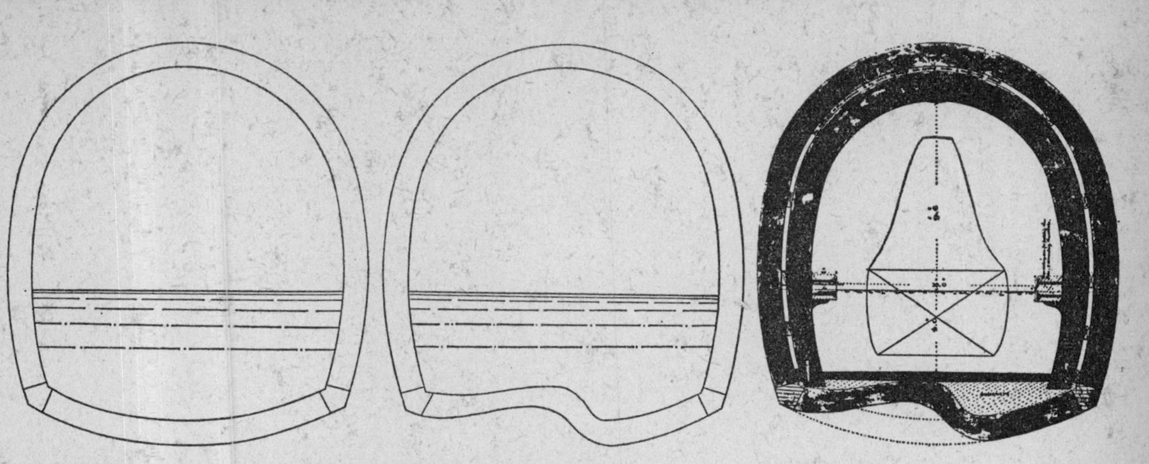

2 (*a*), (*b*) and (*c*). Three drawings to illustrate tunnel movement. The first shows Blisworth tunnel as constructed. The second records an upward movement of a portion of the invert, or tunnel bottom, in 1849. The third shows a temporary repair made by John Lake, the engineer in charge. He erected a series of T-section 15-in × 12-in cast-iron ribs 6 ft apart, each rib being in two halves bolted together at the soffit of the arch, strutted at 4 ft under top water level with a wedged cast-iron stretcher, and fitted with 15-in × 15-in timber fenders at water level. The space between the tunnel brickwork and the ribs above water level were close boarded and the area below the underwater strut was concreted. This repair kept traffic moving until the next annual stoppage of the canal, when the tunnel could be drained, the invert failure removed, and the brickwork reinstated to its original profile.

abroad, powered operation of the paddles and gates has replaced manual working.

Two uncommon kinds of structure were sometimes built on canals: inclined planes and lifts. If a canal were to be constructed over very hilly country, then it would need great flights of locks, expensive to build, slow to use, and requiring large amounts of water. The alternative was to transport boats mechanically from one level to another. One way was to give them wheels underneath, or to put them on trucks, or into tanks, and then haul them on rails by steam or water power from one level to another; these were inclined planes. Another was to raise them vertically, usually in tanks; these were lifts. There is no active inclined plane in Britain now, and only one lift at Anderton, which raises boats 50 ft between the River Weaver and the Trent & Mersey Canal. But in their time they had importance.

Inclined planes were first built in the British Isles by Ducart on the Tyrone Canal in Ireland. Between 1768 and 1773 he seems to have built three timber ramps fitted with rollers up which one-ton boats would be hauled, probably based on earlier Dutch examples. These did not work properly. William Jessop, then working for Smeaton, inspected them, after which they were rebuilt with metal rails, on which larger boats were carried in tanks on trucks, power being by counter-balance; that is, loaded boats going down pulled empty ones up; if necessary the down-going boat could be filled with water.

These Tyrone planes did not last long, but others were built in England, at Ketley in Shropshire in 1788, followed by some on the east Shropshire canals, a number in Somerset, Devon and Cornwall, and others in South Wales. These had all been built by 1838, but later two much larger ones were constructed, at Blackhill on the Monkland Canal near Glasgow in 1850, and at Foxton on the Leicester line in 1900.

Blackhill was erected on a heavily-used coal-carrying canal to save some of the water used to work barges through two parallel flights of eight locks each, arranged in four staircase pairs, each pair with a 24-ft lift. The plane was 1,040 ft long, rising 96 ft mostly at 1 in 10. Two caissons, each 70 ft × 13 ft 4 ins × 2 ft 9 ins almost balanced each other, extra power being given by a 25-hp steam engine. Empty boats used it, loaded ones continuing to pass the locks. Nevertheless, in 1851, 5,452 craft worked over it, and it saved 60 million cu ft of water. Foxton, part of an attempt to widen the Leicester line of canal from narrow boat to barge size, had a rise of 75 ft 2 ins, but slightly bigger caissons. Blackhill closed in 1887, Foxton in 1910. In 1921 the last working plane in Britain, that at Trench on the former Shrewsbury Canal, closed.

Abroad, planes were built on a number of canals, the most spectacular examples perhaps being on the Morris Canal in the United States. This coal and iron carrying waterway, 102 miles long, handling almost 900,000 tons of traffic in 1866, used 23 water turbine-powered inclined planes and the same number of locks to climb 914 ft to its summit level from Jersey City on the Hudson river, and then fall again to the Delaware. The biggest plane had a rise of 100 ft, the smallest of 35 ft; gradients varied from 1 in 10 to 1 in 20. Boats, built in two separate sections, were divided to pass the planes, each half being carried dry, lashed to a cradle.

There was until recently no working example in western Europe, but two new ones have now been finished, a monster to take 1,350-ton barges at Ronquières on the Brussels-Charleroi Canal in Belgium, and another for 350-ton craft at Arzviller on the Canal de la Marne au Rhin in France.

Canal lifts seem to have been first built in England. The most practical examples worked by balancing two tanks of water connected together by cables passing over central wheels. When a small boat needed to be moved from one level to the other, just enough water was added to the uppermost tank to start it downwards, its movement then being controlled by a brake. Seven of this type worked for some thirty years from 1838 on the Grand Western Canal in Devon and Somerset. Much later, the Anderton lift was built in 1875 to carry two narrow boats at a time in each tank. This was partially counter-balanced, partially hydraulically operated. A hydraulic cylinder and ram were placed beneath each tank or cistern. These were connected, and as the weight of water in the uppermost tank forced it downwards, so hydraulic pressure pushed the lower one upwards. In 1908, however, it was converted so that each of the caissons could be separately worked by electric power. The older design, however, had by then been adopted for French and Belgian lifts: the latter are still working. Two other hydraulic lifts were built on the Trent Canal in Canada.

An experimental lift using a different method, floats beneath the tank which would rise and push it upwards when water was added to the float chambers, was built in Britain in the 1790s. The idea was not followed up, but on the Continent two working lifts at Henrichenburg near Dortmund in West Germany, and two others in East Germany, are constructed on this principle.

Water could be supplied to the canal by directly taking in small streams; by making small canals or feeders to obtain it from a river, in which case a weir across the river would be necessary; from reservoirs; by providing a water-wheel or steam-engine to pump from a river, well or colliery below canal level; or by pumping water already used in lockage back from a lower to a higher level.

Feeders would run for miles, such as that from Rudyard Lake, reservoir for the Trent & Mersey Canal, to the former Leek branch, or that which, from its intake off the Taff, ran down beside the river to cross it on an aqueduct before entering the Glamorganshire Canal at Navigation House.

Steam plant, pumping back water from lower levels, was extensively used on the Birmingham Canal Navigations, and was supplemented by water pumped into the canals from mine workings. At Gloucester, steam pumps fed the Gloucester & Sharpness Ship Canal from the Severn; at Crofton on the Kennet & Avon they supplied the canal from springs, and so at many other points on the system. Such pumps might also be necessary to transfer water from reservoirs to the canal itself. Today, diesel or electric pumps are used for similar work. Pumps driven by water-wheels were less common; there is one at Claverton near Bath on the Kennet & Avon, and there used to be another at Welshpool on the old Montgomeryshire Canal.

Water from streams and feeders enters the summit or lower pounds of a canal, some of it working downwards by the use of the locks. The surplus usually passes round the locks in open or covered channels, so keeping all pounds full, but should a sudden storm raise a pound towards the danger level when it might overflow the banks, weirs are provided at intervals to take the excess into the nearest stream. Planks fitting in grooves at narrowings of the canal, or stop-gates, enable sections of the canal to be isolated in case of a burst.

In the canal age, water power drove mills and factories of all kinds. Canal and river navigation companies, to get their Acts, often had to submit to restrictions on their ability to take water from streams that powered industry, and were then relentlessly watched to see these were not infringed. The trans-Pennine canals were on the whole limited to collecting flood-water in reservoirs and, if they used a river as a feeder, to putting in at the top as much as they drew out lower down, the amounts being carefully gauged. The history of such a canal as the Huddersfield is a record of a struggle for water between powerful industrialists who depended on it, and a canal company whose ability to increase traffic lay in getting more of it. Only as works went over to coal did the pressure cease, or, as sometimes happened with flour mills, when the canal company bought them rather than put up any longer with millers being difficult over their water rights.

The canal bed dug and filled with water, the towpath gravelled, bridges, locks, tunnels and other structures built, lock-houses, wharves, warehouses, towing-horse stables erected, all that was now needed was to plant the towpath hedge with quicksets, put up milestones, important because charges for using the canal were by the mile, and tidy up the result of so much effort.

> The Spoil Banks on the whole length of the Canal should be neatly trimmed, and covered with Soil (where it has been saved for the purpose) and early in the Spring they should be sown with Hay Seeds so that they may become returnable land as soon as may be

as the committee of the Nottingham Canal instructed their engineers in 1793. Often trees were planted, for instance on the heaps of spoil round the shafts of the Sapperton tunnel on the Thames & Severn, or in the cuttings of the Welsh Canal, with the idea that later a little extra money might be earned that way.

And so to the ceremonial opening, as of the Bude Canal in Cornwall.

> The Committee of Management, supported by the neighbouring gentry, on the arrival of the loaded boats at the point of debarkation, marched into the town of Holsworthy in procession, the band playing, 'See the conquering hero comes', and hailed by the acclamations of the populace . . . the dinner, provided at the *Stanhope Arms*, was composed of the choicest viands; and the hilarity, happiness and unanimity of all present, were most auspicious.

And now for business.

CHAPTER 6

Carrying Goods

CANALS were built to carry bulk commodities: coal, iron-stone, iron products, other ores and metals, stone for road-making, limestone for the iron manufacturer, or for burning in lime-kilns with slack coal to make lime for spreading on the land, sea-sand containing lime for the same purpose, bricks, timber, slates and other building materials, salt, cotton, wool, corn and flour. Far the most important was coal, replacing wood as a domestic fuel and water as a source of power, demanded to drive the steam-engines Boulton & Watt and other engine-builders were making, to feed furnaces, supply gasworks, warm townspeople, and fulfil the other needs of an expanding economy.

A trade in finished goods such as textiles or beer could be built up, or in imported groceries such as sugar, and some canals also carried passengers. But in their heyday, British canals were bulk carriers, as they are still today. The same is true abroad: coal, oil, bulk chemicals, ores and metal products, grain, form the basic cargoes of the world's waterways, though other trades have been built up, like the carriage of finished motor-cars on some big American water channels.

Cargoes could reach the canals by a colliery or works branch, by horse tramroad, or by road waggon.

The Duke of Bridgewater was exceptional in being able to carry coal by boat from within his mines direct to Castlefield basin in Manchester, whence it was distributed to customers. He used small containers, loaded at the coal-face, placed in the boats, and lifted out again on to coal carts. But many collieries had private canal branches and basins where boats lay to load.

So did quarries. The curious can follow the Thanet branch of the Leeds & Liverpool Canal behind Skipton to the point where barges were loaded from the quarry shoots.

We do not perhaps now realize that, in the canal age, some 1,500 route miles of horse tramroads fed the canals with loads from mines, quarries and works, some many miles from the waterway which carried their goods. The Derby Canal company itself owned a tramroad, down which containers of coal came from the collieries in two-ton waggons. They were transferred to barges at Little Eaton, carried by water to coal merchants' wharves, and then re-transferred by crane to carts for delivery. Again, limestone from the great Caldon Low quarries came steeply down a tramroad with its own inclined planes to the Caldon branch of the Trent & Mersey Canal at Froghall. Here some was burnt into lime at the row of great kilns on the wharf, with coal slack brought up the branch by boat, and the rest was boated away, mainly to the Staffordshire ironworks to be used in iron-making.

Bulk cargoes also reached the waterways by road waggons. They could expect a reasonable surface on the turnpikes, though they did not always get it, as when the coal trade was reported seriously hampered by the impassable state of the Market Harborough & Kettering turnpike in 1818, but side roads leading to wharves were often maintained by the parish, who sometimes tried and failed, and sometimes did not try, to cope with the heavy wear these roads got. A canal company might try the effect of threatening indictment of the parish authorities before Quarter Sessions, as the Grantham Canal company did for the road from Harby towards Melton Mowbray in 1800 and 1804. Or they might offer to help pay, as the Leicestershire & Northamptonshire Union company did for two roads in 1797, in one case providing gravel if the parish found the labour, in the other by agreeing to maintain it themselves if the parish paid a cash contribution. Because local roads were so often damaged by subsidence due to salt workings, the Weaver Navigation, which carried much of the salt produced, frequently subsidized repairs, and sometimes paid the whole cost, but then it was a county body. In addition to such special help, many Acts compelled canal companies to carry stone for use on neighbouring roads either toll-free, or free as long as they did not use water by passing through a lock when the water was not running over the weirs.

Works, receiving raw materials or sending away finished products, often had a side-arm off the canal, with a neat hump-backed bridge over it if on the towpath side. Others built warehouses, perhaps with overhanging canopies beneath which the boats could unload in shelter, or with an arched entrance for a canal branch to run inside. Occasionally unloading was done from a warehouse built right over the canal.

All these were private. In addition the canal company would itself build walled-in terminal and intermediate basins at important towns, with wharves, cranes and warehouses, in charge of their own wharfinger. These were public, and any carrier could use them on payment of wharfage, cranage and warehouse charges. The Rochdale Company were proud of their busy wharves, and in 1822 ordered

> that a Gateway built of stone rusticated, be erected at the Entrance into the Manchester Wharf.

It is still there, forming, rather sadly, the gate to a car park. At a big centre, especially if it were also a transhipment point for the transfer of cargoes from narrow boats to barges, like Stourport, or boats to river or seagoing craft, like Goole, Runcorn or Ellesmere Port before the Manchester Ship Canal was built, there might be specialized wharves for particular commodities like iron or pottery clay, and warehouses for corn, salt, cheese or the special products of the area.

Other wharves and warehouses, with counting-houses included, were owned or rented by carrying firms. Canals were thought of as water roads, the companies maintaining the track, on which anyone who wished could put boats. A carrier might be a couple of men running a single boat, or a firm with

At a General Assembly of the Company of Proprietors of the GRAND JUNCTION CANAL, *held at the Company's Office,* 21, *Surry Street, Strand, London, in the County of Middlesex, on Tuesday, the Second Day of December, One Thousand Eight Hundred and Fifty One, it was*

RESOLVED AND ORDERED—

THAT Boats trading on the Grand Junction Canal having relays of horses, which shall arrive at a Lock during the passage through the same of Boats or Barges not having relays of horses, shall be entitled to pass the said Boats or Barges in the top or bottom Pound of the said Lock; and any Persons having charge of Boats or Barges not having relays of horses, who shall refuse, when called upon by any servant of the Company, to stop their Boats or Barges when they shall have passed through the Lock, shall forfeit and pay for the first offence the Sum of TWENTY SHILLINGS, and for every subsequent offence the Sum of FORTY SHILLINGS.

CHARLES ROGERS,

Clerk to the said Company.

3 The Grand Junction orders priority at locks for fly-boats, which worked with relays of horses

two or three dozen boats, employing the boatmen, owning the towing horses, and carrying on special contract or on scheduled runs. The latter could be stage-boats for heavy goods, or fly-boats, carrying double crews, using relays of horses, and working round the clock, for fast consignments and parcels.

In 1825, it was said that stage-boats were taking three to four days for the 93-mile run from Birmingham via Fazeley to Preston Brook, where the Trent & Mersey met the Bridgewater Canal, and cargoes were transhipped to bigger craft for the run down the Duke's canal to Runcorn, and then down the estuary of the Mersey to Liverpool. Fly-boats took 44 hours, a speed of just over 2 mph non-stop, day and night, through 96 locks, and the tunnels at Harecastle, Barnton, Saltersford and Preston Brook. There were probably more independent men owning their own boats on the navigable rivers than on the canals. For instance, most of the Humber boatmen were independent and some bought and sold cargoes as well as acting as carriers. But on the canals they were always a small minority.

The best known of the carrying firms is Pickfords. They started as road carriers in the mid-eighteenth century, and do not seem to have added canal carrying to their waggon business until about 1790. For the next fifty years they operated a concern of steadily increasing size and financial shakiness. They worked especially on the Manchester to London run via the Grand Junction as soon as that route opened in 1800, on the trades via the Staffs & Worcs Canal to the Severn at Stourport, and so down to Bristol, and on the Leicester line of waterways.

Here is a typical Pickfords advertisement, from *Aris's Birmingham Gazette* of November 28th, 1814.

EXPEDITIOUS WATER CONVEYANCE FROM BIRMINGHAM TO LEICESTER, &C.,

MESSRS. PICKFORD beg leave to inform their Friends and the Public in general, that they have established a Pair of FLY

STAGE BOATS weekly from hence to Leicester, and intermediate Places, which load goods at Birmingham every Thursday Afternoon, Warwick every Friday, Banbury and Oxford every Saturday; and discharge at Market Harborough and Leicester every Monday, return from Leicester every Monday evening; discharge Warwick Goods every Wednesday; and Birmingham every Thursday morning.
By these Boats Goods are regularly conveyed to and from all Wharfs and Places on the Line of the Birmingham, Warwick and Birmingham, Warwick and Napton, and Oxford Canals.
An Arrangement is making, and will soon be completed (of which timely Notice will be given) for extra Boats to leave Birmingham every Tuesday.

RATES TO LEICESTER.	Light Goods	2/6d per Cwt
	Heavy ditto	2/3d ditto
	Tyre, Iron and Nails	2/0d ditto

Messrs. PICKFORD'S Fly Boats continue to load daily for London and intermediate Places, as usual, and every attention is paid in forwarding and delivering Goods with the greatest Regularity and Dispatch. Rates of Freight &c., may be had by applying to their Agent, Mr. Joseph Hunt, Warwick Junction Wharf, Bottom of Fazeley Street.
Birmingham, November 7th, 1814.

Of the other firms, Worthington & Gilbert, in which John Gilbert, the son of the Duke's agent, was a partner, had a privileged position on the Duke's canal. The Anderton Carrying Co specialized in carrying goods to and from the Potteries by way of the Trent & Mersey and Weaver, transhipping at Anderton. Samuel Bache, Thomas Sherratt, Skey, Small & Co, Samuel Danks, G. Wheatcroft & Sons, and, over the Pennines, John Thompson, James Veevers, and the Rochdale and Halifax Merchants Co, are a few of many. Poor Thomas Sherratt: he owned his own business, went bankrupt, retained the managership of the fly-boat section, and then in 1814 advertised that every day at 10 AM he was available in Birmingham to impart useful information on road and canal carrying

> 'obtained in an Experience of nearly 50 years'

at 3*s* 6*d* an hour. Let us hope that some came to consult the old gentleman, in spite of his failure.

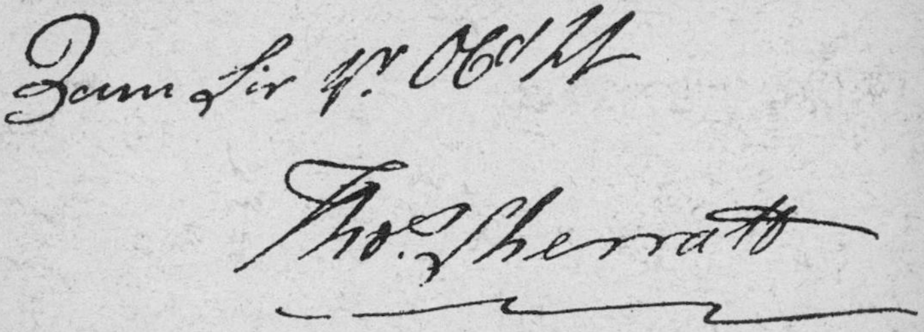

4 Thomas Sherratt's signature in 1798

Advertisements in *Aris's Birmingham Gazette* give the feel of the times:

March 25th, 1811: Crowley, Hicklin & Co. run fly-boats to London and stage boats to Kidderminster and Stourport every Thursday, loading there on Mondays to return. They provide lock-up boats for wine and spirits, and if required send goods by land (at the higher land rates) when the canals arc stopped by frost.

April 22nd, 1811: William Judd's[1] boats for Southampton, Portsmouth and Gosport loaded every Monday and Thursday, goods being delivered in nine days.

[1] Sherratt was his manager.

These would probably have gone by water to London, possibly to Basingstoke, and then by road. Judd also had boats to Banbury and Oxford.

July 5th 1813: Skey, Small & Co's boats load at Birmingham and Wolverhampton 'every Spring Tides, three days before the Full and Change of the Moon, for Bristol, Gloucester, and all western parts of England'.

Not till the Gloucester & Sharpness Canal was opened in 1827 to by-pass part of the Severn, was trade freed from dependence upon the spring tides to find enough water.

July 4th, 1814: J. Jackson and S. George send fly-boats to London twice a week, and despatch regularly to Tamworth, Nuneaton, Hinckley, Leicester and Coventry.

Goods for Leicester would be taken by canal to Hinckley on the Ashby Canal for onforwarding by road waggon.

Sept. 12th, 1814: Thomas Coleman loads daily for Liverpool, Manchester & Chester. From Liverpool goods can be shipped to Lancashire and Cumberland ports, Glasgow, Ireland or North Wales. Boats also load for Gainsborough and Hull for Sheffield, Lincolnshire, the East and West Ridings, Newcastle, Edinburgh and eastern Scotland.

Many carriers' businesses were substantial. William Jackson & Son, operating on the Rochdale Canal, in 1845 rented warehouse space at Manchester, Rochdale and Sowerby Bridge from the canal company. The concern then had 32 broad boats (and 2 building) and 6 narrow, worth an average of £230–£240 each, and 112–20 horses. They valued their stock at £12,000.

Until railway competition began, not many canal companies themselves engaged in carrying, preferring to confine themselves to providing the track, except sometimes when a canal had just opened, when the company itself ran boats to encourage a trade to get going, but dropped out as soon as possible. But there were exceptions; notable among them were the Trent & Mersey, which ran an extensive carrying business of its own through a subsidiary, Hugh Henshall & Co, whose headquarters were in the canal company's offices at Stone; the Mersey & Irwell, operating a large fleet, mainly of sailing flats, between Manchester and Liverpool, and the Duke's canal, which first worked through a linked carrying concern, Worthington & Gilbert, and later, directly by the Bridgewater Trustees' carrying fleet. All these operated on their neighbours' waterways as well as their own.

Very large fleets were, however, owned and operated by colliery owners, rock-salt mine proprietors and others who produced the bulk goods to be carried. On those waterways which had been built with the carriage of one or two commodities especially in mind – salt and coal on the Weaver, coal and iron on the Shropshire or the South Wales canals – most of the craft were producer-owned and managed.

Cargoes were charged for by the ton carried a mile. Maxima for various classes of goods were laid down in the canal company's Act – sometimes amended later – but below these, changes could be made by shareholders' meetings.

The promoters of the first canals were considered to be providing a much-needed facility, and Parliament was anxious to encourage them to risk their money. Therefore generous maxima were granted. This continued to be the case for later canals which were needed to serve a colliery or manufacturing area – such as the Cromford Canal to Butterley ironworks, collieries and limestone quarries on the line, and the textile mills at Cromford, or one to carry coal and other necessary commodities into a very rural area, and which, like the Oakham Canal into Rutland, could not expect a large trade.

But as the bigger canals, like the Trent & Mersey, began to show how profitable they could be, so investors no longer had to be persuaded to risk their money; they offered it, and in the canal mania of 1793 threw it at those busy forming canal companies, with good, poor, or hopeless prospects alike. Those who wanted their goods carried, and those who would benefit from the trade, therefore encouraged Parliament to lower the maxima as far as possible without actually preventing the investment. So, whereas the Trent & Mersey in 1766 had been granted tolls of 1½*d* a ton on everything, the Grand Junction in 1793 was given 1*d* on merchandise, ¾*d* on coal, ½*d* on building material, and ¼*d* on lime and limestone. But these were maxima: competition between canal companies, or rival routes, or between canals and road transport or coastal shipping, often kept actual tolls well below those authorized.

Occasionally other forms of charge were made, such as an annual licence fee, sometimes used on Welsh canals when boats were owned by a colliery or similar company and extensively engaged in carrying that company's products, and, usually, also for fly-boats carrying light goods and parcels.

Credit was usually up to three months for approved accounts with carriers or producers; occasionally up to six. Others paid cash, boatmen being given what they would need to pay tolls along their route. This sometimes caused difficulties in the days of small country banks and many failures. The Nottingham Canal company in 1809 told its toll-collectors to take only cash, Bank of England notes, or

> the Notes of some of the Bankers in the Counties of Notts, Derby, Leics and Lincoln, and no other.

The Leicester company in 1814 observed sadly that of the receipts for June of £365, £34 was in the notes of banks that had stopped payment.

BIRMINGHAM CANAL.

No. 7829 Loaded 30 Sepr 1790
Boat No. 4 W Belker Steerer.
M Pickerd Owner.
Loaded at Mr Astons Miles.
To Brmm 11¼ Ditto.
Ditto.

T.	C.	Q.			£.	S.	D.
17	0	0	Coals	Miles.	1	3	10½
			Coaks				
18	10						

Hen Wright
J. Dawkes

John Houghton Collector.
J Bentley Overseer.
The above Contents are true.

5 A Birmingham Canal toll-ticket of September 30th, 1790, for the carriage of 17 tons of coal from Mr Aston's wharf for 11¼ miles to Birmingham. The toll was 1½*d* per ton per mile. It looks as if the boat was gauged by Messrs Wright and Dawkes, and found to be carrying 18½ tons

As between pike-keepers and carriers on the turnpike roads, so between toll-collectors and boatmen on the canal, a perpetual battle of wits went on. At first boatmen carried a waybill issued at the loading point by the consignor of the cargo saying what tonnage of each kind of commodity was in the boat,

FOWLER's and Co. *Birch-Coppice* Coal

Tons. Cwts.

Coal delivered 18 „ 0 Bradley Green Wharf

To Boat, No. 1. Coventry Co.

For Tho Homer & Co

Steered by James Patchel

15 *Day of* 5 mo 1792

E„d Tho Rathbone

6 A waybill issued by a colliery company

where it had been loaded, and its destination. The first toll-office they passed calculated the tolls on that company's canal, and collected the cash. Boatmen soon started to load more cargo than the waybill stated, hoping to get the excess through without payment. The company's answer was weighing and gauging.

Each boat had to be taken to a company dock to be indexed. The gunwale when empty was marked. Weights of about 2½ cwt each were then put in, two at a time, by cranes, the distance from the first mark to the water level being measured, and recorded each time. These readings were entered into books issued to each toll-collector, who therefore had the vital statistics of every boat moving on the waterway. To check or gauge the tonnage carried, he had only to measure with a gauging-rod from the gunwale mark to the water level, and consult that reading in his book. Some companies, instead of gauging the craft, compelled them to carry index plates, at bow and stern, which showed the tonnage carried at the depths marked. These readings were then averaged to give an acceptable figure.

More difficult to detect were boatmen who, with the connivance of the consignor, hid a few tons of a commodity carrying a high toll under a cargo of limestone or sand. If any such were caught, they were usually prosecuted and their credit stopped. Again, there could be collusion between toll-keepers and boatmen, the risk being minimized by asking for good references and security bonds from toll-keepers, and not allowing them to have any side business that might get them in the boatmen's power.

Apart from the tolls paid to the canal company for the use of the track, the carrier had also to be paid for the services of his boat, horses and men; he would normally quote an inclusive charge to a customer, covering both freight and tolls.

There was no legal control over carriers' charges. When business was good, they added as much as they dared; when it was bad, they took it off. They could also vary their charges for each of their employers, whereas canal companies, until 1845, had to charge all comers the same toll for similar cargoes over the same distance. It was indeed high freight charges, probably more than high canal tolls, which so much encouraged the development of railways.

Because carrying was usually done by independents, canal companies had nothing corresponding to a commercial department seeking business. When a request for a change of toll came in, perhaps to enable a particular commodity to develop a new market, or for a drawback, that is, a reduction of toll beyond a certain distance conveyed, from a carrier or a producer, they very often responded, but very seldom did they take the initiative, and then usually because of the private contacts of members of the committee. Those companies with

No. 1046, late No. 35 reweighed December 3rd, 1818.

—o❖—o❖o—❖o—o❖o—o❖o—

THIS Boat was built by Mr. Samuel Barnsdall of Newark, in the year 1797, for Mr. John Derry of Clifton Hill. She has been in the hands of several Proprietors. The present Owner purchased her of Mr. John Mountain of Boston, in the year 1818, and has chiefly employed her in the Coal Trade to Boston and Donnington Bridge, and in bringing Corn back to Nottingham &c.

This Boat had been trimmed about twelve months before these gauges were taken. Part of her Floor was taken up, and she was found rather *foul* and *leaky*. Her length is 68 feet, and breadth, across the Mid-ships 13 feet 11 inches. She drew 10 inches Water when light, and 33. 42 inches when laden with 40 Tons.

When these gauges were taken there were on board,—a Mast, Sail, and complete standing Rigging, Blocks, Lines, &c. six Poles, one Oar, two Deal Planks, one Windrope, one Headfast, one Sternfast, one Cadger, one Iron Crow, one Corn-tub, an immoveable Cabin, and a Firestand; also seven Covers and Dunnage, that were not on board, that weighed 12¾ Cwt. and seven Bundles of Bags that weighed 4 Cwt. She had Side Cloths, that were fastened to her Gunnels all round.

N. B. The length of the Fore-hold is 18 feet 11 inches, and Hinder-hold 22 feet 8 inches.

7 (*a*) and (*b*). Facing pages from one of the printed volumes of gauge tables issued by the Trent Navigation on behalf of some ten East Midlands companies. Each pair of pages records one boat: the left-hand page describes her in detail as she was when she was gauged; the right-hand page shows how much freeboard she will have when carrying cargoes up to 51 tons, eg, if the freeboard is 16·41 ins, she will be carrying 30 tons. By measuring her freeboard, therefore, the tonnage carried on any trip can quickly be found by any toll-keeper who possesses a set of printed tables

Robt. Hill, Donnington Bridge, No. 1.—Rt. Glover, Master.

Tons.	Dry Inches.	Difference.	Tons.	Dry Inches.	Difference.	Tons.	Dry Inches.	Difference.
Light	34. 28	.66	19	22. 76	.58	38	11. 95	.55
1	33. 62	.65	20	22. 18	.58	39	11. 40	.54
2	32. 97	.64	21	21. 60	.58	40	10. 86	.54
3	32. 33	.64	22	21. 02	.58	41	10. 32	.54
4	31. 69	.63	23	20. 44	.58	42	9. 78	.54
5	31. 06	.62	24	19. 86	.58	43	9. 24	.53
6	30. 44	.61	25	19. 28	.58	44	8. 71	.53
7	29. 83	.61	26	18. 70	.58	45	8. 18	.53
8	29. 22	.60	27	18. 12	.57	46	7. 65	.53
9	28. 62	.59	28	17. 55	.57	47	7. 12	.52
10	28. 03	.59	29	16. 98	.57	48	6. 60	.52
11	27. 44	.59	30	16. 41	.57	49	6. 08	.52
12	26. 85	.59	31	15. 84	.56	50	5. 56	.52
13	26. 26	.59	32	15. 28	.56	51	5. 04	
14	25. 67	.59	33	14. 72	.56			
15	25. 08	.58	34	14. 16	.56			
16	24. 50	.58	35	13. 60	.55			
17	23. 92	.58	36	13. 05	.55			
18	23. 34	68	37	12. 50	55			

As 51 Tons put this Boat down 29.24 Inches, one Ton, upon an average, puts her down .57 of an Inch.

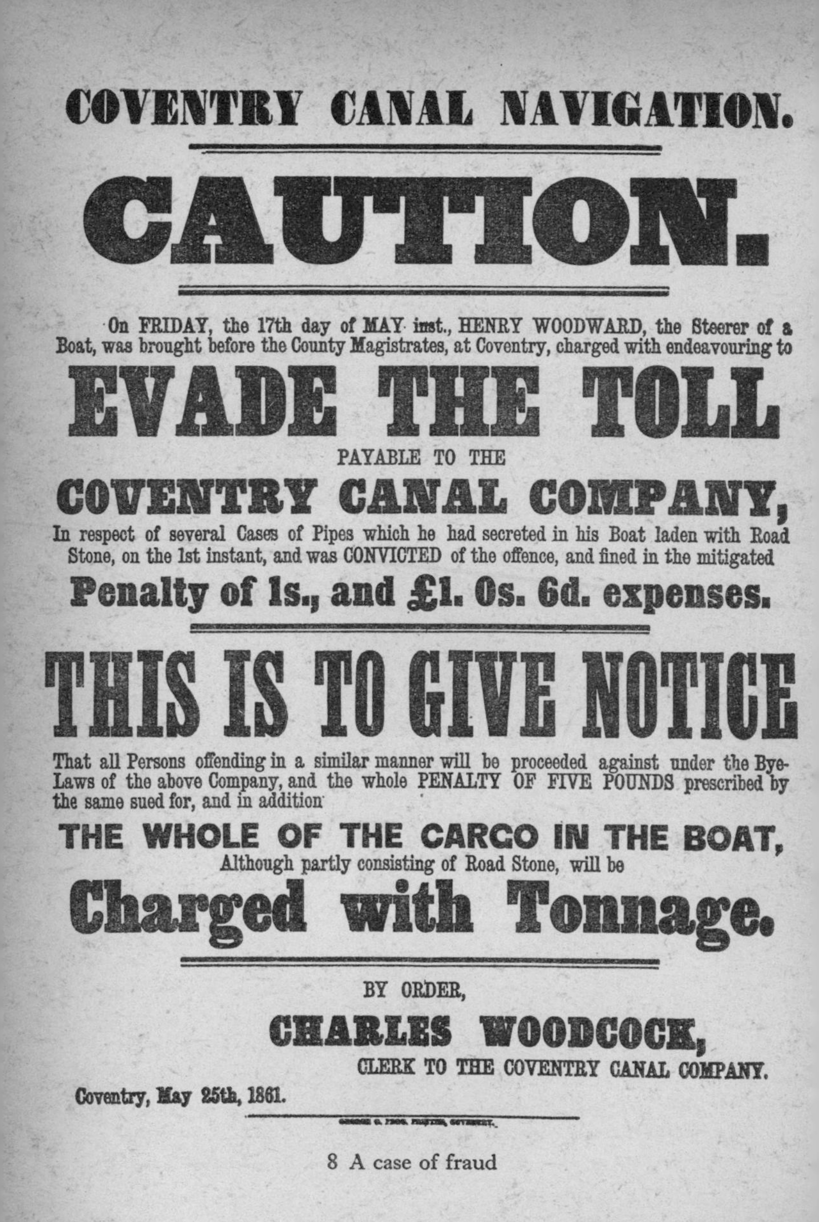

COVENTRY CANAL NAVIGATION.

CAUTION.

On FRIDAY, the 17th day of MAY inst., HENRY WOODWARD, the Steerer of a Boat, was brought before the County Magistrates, at Coventry, charged with endeavouring to

EVADE THE TOLL

PAYABLE TO THE

COVENTRY CANAL COMPANY,

In respect of several Cases of Pipes which he had secreted in his Boat laden with Road Stone, on the 1st instant, and was CONVICTED of the offence, and fined in the mitigated

Penalty of 1s., and £1. 0s. 6d. expenses.

THIS IS TO GIVE NOTICE

That all Persons offending in a similar manner will be proceeded against under the Bye-Laws of the above Company, and the whole PENALTY OF FIVE POUNDS prescribed by the same sued for, and in addition

THE WHOLE OF THE CARGO IN THE BOAT,

Although partly consisting of Road Stone, will be

Charged with Tonnage.

BY ORDER,

CHARLES WOODCOCK,

CLERK TO THE COVENTRY CANAL COMPANY.

Coventry, May 25th, 1861.

8 A case of fraud

USEFUL and CORRECT

ACCOUNTS

OF THE

NAVIGATION,

OF THE

RIVERS AND CANALS

West of London.

COMPRISING, IMPORTANT AND INTERESTING

PARTICULARS of INFORMATION;

With Tables of Distances; Time of Navigating; and

PRICES of CARRIAGE,

ON EACH RIVER and CANAL.

PREPARED FOR THE USE OF

RIVER and CANAL SHARE PROPRIETORS, COMMITTEES, MERCHANTS, CLERKS, AGENTS, MANUFACTURERS, WHARFINGERS, NAVIGATORS, and TRADERS IN GENERAL.

The whole illustrated with

A NEAT ENGRAVED and COLOURED MAP.

The Second Edition, much improved,

BY Mr Z. ALLNUTT, HENLEY,

Superintendant and Receiver &c. on the Thames Navigation.

PRINTED FOR THE AUTHOR AT HENLEY:

SOLD BY Mr ASPERNE, 32 CORNHILL: Mr TAYLOR 59 HOLBORN: Mr FULLER 61 CHARING CROSS and Mr GIFFORD 158 STRAND, LONDON.

9 The title-page of Zachary Allnutt's handbook

their own carrying businesses were much more immediately responsive to demand. They also had the advantage that as carriers they could vary their freight charges to get or keep business, whereas as canal owners they could not do so.

Zachary Allnutt, superintendent and receiver of the tolls for the Thames Commissioners in the 1800s, was, however, a commercial manager *manqué*, for in 1810 he prepared and published, seemingly at his own expense, a handbook for

> river and canal share proprietors, committees, merchants, clerks, agents, manufacturers, wharfingers, navigators and traders in general,

which described the facilities not only of his own river navigation, but of others that were accessible from it, as far away as the Gloucester & Sharpness Ship Canal, the River Wye, and the canals and horse tramroads of Wales. It contained a map, distances, opening hours, and useful information on wharves, commodities carried, barge sizes etc, and sold for 3*s* 0*d* in paper, 3*s* 6*d* interleaved, or 4*s* 0*d* with pasteboard covers. He gave both tolls and specimen freight charges, and compared them, favourably, to the road carriage charges he also quoted.

He loved his subject, and warms us at once with his opening lines on his own Thames:

> This noble and beautiful River, acknowledged to be of the first importance for Trade, not only in Great Britain but in the whole World. . . .

Yet it is admirably concise, and authors would do well if they could say with Zachary:

> The Editor of these Accounts of River and Canal Navigations, begs to take this opportunity of stating to the Public, that the many Particulars herein detailed (which have been obtained from late Surveys, authentic Documents, and the most minute researches), are compiled with the utmost care and attention to their accuracy, brevity and perspicuity.

CHAPTER 7

Carrying Passengers

NOWADAYS, people only travel on business through the ship canals, but many an inland waterway carries the pleasure seeker and the explorer: the steamer *Wilhelm Tham* and his brothers of the Göta Canal, smaller passenger craft on such Scandinavian canals as the Dalslands and Kinda in Sweden and the Telemark in Norway; the Köln-Düsseldorfer Rhine boats and those of other companies on the same run; the craft that tour the Dutch bulbfields, or the hydrofoils of the Danube. And every year the waterways of Britain, North America and, to a lesser extent, the Continent see more and more pleasure craft, owned or hired, their crews discovering the many pleasures that only canals can give.

But to our forefathers, inland waterways were sometimes the best, sometimes the only, sometimes the cheapest, means of passenger transport. In England they played a comparatively minor part, for our canal building coincided with the setting up of turnpike trusts to improve the main roads, and the growth of stage- and later mail-coach services on them, running at increasing speeds.

Many had market boats, picking up passengers and goods early in the morning, taking them to the local market town, and bringing them back again in the evening. Some had longer-distance services. The Duke of Bridgewater began it in 1772, with two boats between Manchester and London bridge near Warrington:

> one carried six score passengers, the other eighty: each boat has a coffee-room at the head, from whence wines, &c., are

sold out by the captain's wife. Next to this is the first cabbin, which is 2*s* 6*d*, the second cabbin is 1*s* 6*d* and the third cabbin 1*s* for the passage.

The Duke's boats were so successful that regular services were established on a number of waterways, often with coaching connexions: Manchester and Runcorn (with connexions to Liverpool); Manchester and Worsley, later extended to Leigh, Wigan and Scarisbrick (for Southport); Manchester and Bolton; Manchester and Ashton-under-Lyne; Bolton and Bury; Liverpool and Wigan; Preston and Kendal; Selby and Hull; Knottingley, Goole and Hull; Chester and Ellesmere Port; and in the South, Paddington and Uxbridge and Bath to Bradford-on-Avon. Some services were run by the canal companies, others by lessees, often with craft hired to them, and on condition that certain regulations were observed, such as,

> Refreshments with Wine, Ale, Porter, Cyder or Perry are permitted but no liquors are to be kept, or any smoking suffered on board the Boat,

or,

> The Captain and an assistant to be in the Boat besides the Woman who attends the kitchen, and a person to ride the horses.

Such craft were given priority. The Duke's flourished a sickle on the bow, symbolically threatening to cut the towline of any barge that got in its way. Other packet or passage boats, as they were called, ran under the protection of some such bye-laws as:

> Any Boat obstructing the Passage of the Packet Boat ... shall forfeit Five Shillings,

THE PACKET BOATS

BETWEEN

PRESTON AND LANCASTER,

WILL BEGIN

ON MONDAY THE THIRD SEPTEMBER, 1804,

TO STOP AT

THE CANAL BASIN,

NEAR THE

CHAPEL-YARD, IN PRESTON,

WHERE

They will take in and deliver their Parcels and Passengers.

A SITUATION SO CENTRAL & CONVENIENT

TO THE

TOWN AND TRADE OF PRESTON

It is hoped will prove a great Inducement to public Encouragement, and nothing will be wanting on the Part of the Proprietors to continue a strict Attention to the Convenience of the Passengers, and the safe and expeditious Conveyance of the Parcels intrusted to their Care.

Preston, 30th *August*, 1804.

W. ADDISON, PRINTER, PRESTON.

10 The packet boats on the Lancaster Canal change their terminus at Preston. A notice of 1804

and

> If any Boat navigating the Canal do not drop its Line and give the inner side to the packet boat in passing, the Manager of such Boat shall forfeit 10*s*.

Flourishing services also grew up in Scotland. Boats ran before 1800 from Glasgow to Falkirk, whence passengers were taken on to Edinburgh by coach, but after 1822, when the connecting canal was opened to the capital, through services began, both by day and, with sleeping accommodation, at night. Others ran on the Monkland Canal between Sheepford near Airdrie and Glasgow, and on the Aberdeenshire between Inverurie and Aberdeen. On most of these craft refreshments were served: but on the Inverurie–Aberdeen run the rule was no spirits and no tipping. Apart from regular services, there were also excursion boats, for instance, in the 1770s to Chester races, for which cheap day-return tickets were issued; in the 1790s and 1800s, from Chester to the Mersey for bathing; and later for all sorts of occasions, as from Glasgow to Edinburgh to see King George IV, or the varied trip between the same two cities offered in 1837, and taking seven hours: from Glasgow to Castlecary by canal boat, then by coach to Stirling; to Newhaven by steamer; and then to Edinburgh by coach again.

Or one could organize one's own outing, as did the rector of Camerton in 1823 on the Somersetshire Coal Canal:

> Having engaged one of the coal barges, I had it fitted up for the ladies with an awning and matting against the sides, and tables and chairs from the public-house, in which we proceeded about eleven o'clock to Combe Hay, where we visited the Mansion House, walked round the premises, and afterwards dined under the trees near the cascade. As the day was delightful the whole party much enjoyed the excursion.

Occasionally there might be a mishap, of the kind reported in a newspaper of 1847, that,

> A canal boat containing 170 persons on a holiday excursion to Birmingham sank on Easter Monday in a canal at Smethwick, in 6½ ft of water. Every one escaped miraculously.

The passenger boats were operated at speeds of 3 or 4 mph drawn by two or three horses, one usually ridden by a postilion dressed in uniform, perhaps the black cape and scarlet jacket of the Glasgow, Paisley & Ardrossan, or the yellow cape and blue jacket of the Grand Junction. The Duke tried to get more speed by using mules, but only at risk of some damage to the canal banks, for an observer of 1805 wrote:

> We observed a constant elevation of the water before the passage boat, as it moved along, of at least 9 inches, and perhaps more than a foot at times; and the rapidity with which the water ran backwards, between the boat and the side of the canal, appeared to have a most destructive effect upon the latter, particularly on the towing path side; and often this was laid quite under water, for considerable distances together, by the surge or wave opposite to the head of the boat as it passed along; while the labour of towing was most materially increased.

For a brief time just before the railway age began, William Houston made an outstanding success of fast passenger carrying on the Glasgow, Paisley & Ardrossan Canal between Glasgow, Paisley and Johnstone. He discovered that if the boat were lightly enough built, it could be made to skim the water, creating hardly any wash, and maintaining a speed of 9 to 12 mph. In June 1831 he introduced a passage boat 70 ft long and 5½ ft wide, with a thin iron hull that weighed less than 1¾ tons, and a loaded draught of only 16¾ ins. To save weight, covered cabin and steerage accommodation was provided by stretching oiled cloth over curved ribs, with spaces left for windows. There was also open seating in the bows, and baggage accommodation under the bow and stern decks.

Pulled by two galloping horses as well bred as those in the mail-coaches, and changed every four miles, carrying 80 to 90 passengers each paying 1*s* cabin or 9*d* steerage for the trip from Glasgow to Paisley, these swift boats, easy-riding and fast, drew passengers away from the coaches, 373,290 of them in 1835 and 423,186 in 1836, travelling in twelve boats a day in each direction.

Houston's swift boats were copied on the Glasgow–Edinburgh and Preston–Kendal runs; on the latter, two-horsed boats were put on in 1833, limited to 50 passengers, and taking seven hours for the 57 miles, including stops and changing

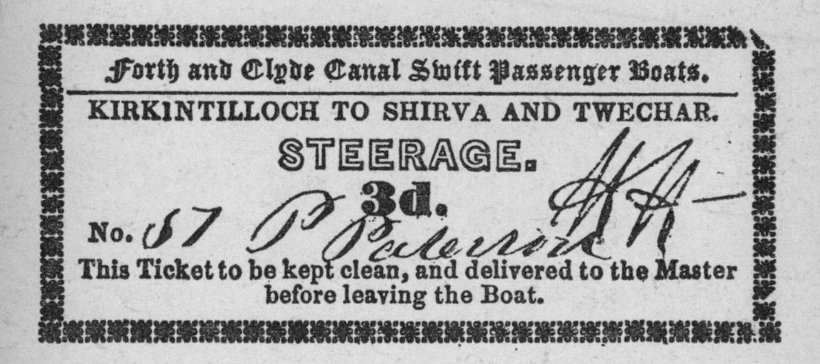
Forth and Clyde Canal Swift Passenger Boats.
KIRKINTILLOCH TO SHIRVA AND TWECHAR.
STEERAGE.
3d.
No. 67
This Ticket to be kept clean, and delivered to the Master before leaving the Boat.

11 A ticket for a swift boat on the Forth & Clyde Canal

horses. By 1839 four boats were being used, with three services a day in summer. On the Duke's canal, swift boats, pulled by two or three horses ridden by a red-coated postilion, the leader with a horn to warn traffic to give way, ran from Manchester at varying hours to catch the ebb tide at Runcorn, where passengers changed for Liverpool.

Thomas Grahame wrote of such boats in 1833: 'They are

more airy, light and comfortable than any coach', and Sir Archibald Geikie in his reminiscences:

> The boats were comfortably fitted up and were drawn by a cavalcade of horses urged forward by postboys. It was a novel and delightful sensation, which I can still recall, to see the fields, trees, cottages and hamlets flit past as if they formed a vast, moving panorama, while one seemed to be sitting absolutely still. For mcre luxury of transportation, such canal travel stands quite unrivalled.

Several English canal companies took an interest in Houston's fast boats. In 1833, they imported one of them, and used it for tests. It finished up running a passenger and parcels service between Bath and Bradford-on-Avon, where it was

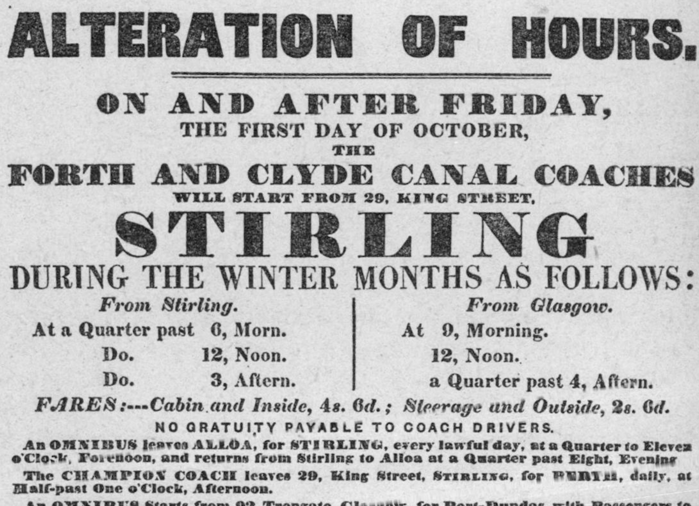

ALTERATION OF HOURS.

ON AND AFTER FRIDAY,
THE FIRST DAY OF OCTOBER,
THE
FORTH AND CLYDE CANAL COACHES
WILL START FROM 29, KING STREET,
STIRLING
DURING THE WINTER MONTHS AS FOLLOWS:

From Stirling.	*From Glasgow.*
At a Quarter past 6, Morn.	At 9, Morning.
Do. 12, Noon.	12, Noon.
Do. 3, Aftern.	a Quarter past 4, Aftern.

FARES:---Cabin and Inside, 4s. 6d.; Steerage and Outside, 2s. 6d.

NO GRATUITY PAYABLE TO COACH DRIVERS.

An OMNIBUS leaves ALLOA, for STIRLING, every lawful day, at a Quarter to Eleven o'Clock, Forenoon, and returns from Stirling to Alloa at a Quarter past Eight, Evening

The CHAMPION COACH leaves 29, King Street, STIRLING, for PERTH, daily, at Half-past One o'Clock, Afternoon.

An OMNIBUS Starts from 92, Trongate, Glasgow, for Port-Dundas, with Passengers to go by the Boats, 25 minutes before the Starting of each Boat, going along Trongate, Argyle Street, and up Buchanan Street; and also waits the arrival of each Boat to take the Passengers to the City.—Fare 3d., and a Penny for each Light Portmanteau, Travelling Bag, &c.

CANAL OFFICE, 28th September, 1841.

12 Coaches from Stirling connect with canal packet boats to Glasgow in 1841, a single fare covering both journeys

always called the Scotch boat. The trip took under 1½ hours. Passengers had the choice of first- or second-class cabins, and were entertained, at any rate sometimes, by a string band.

In 1839 the Lancaster Canal were operating a twice-daily swift-boat service between Kendal and Preston, with a third boat between Lancaster and Preston. One boat, for instance,

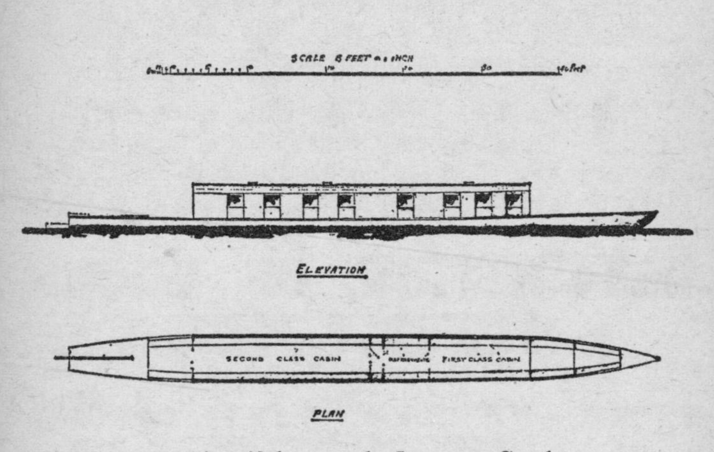

13 A swift boat on the Lancaster Canal

left Kendal at 6.30 AM and covered the 57 miles to Preston in just under 7 hours, arriving at 1.20 PM. It turned round in 10 minutes, leaving again at 1.30 to arrive back at Kendal at 8.50 PM. The single fare was 6*s* first class, or 4*s* second. At Preston, omnibuses worked between the canal wharf and the railway stations, where connecting trains could be got for Liverpool or Manchester, the latter in turn connecting with others for Birmingham. Coaches to Bolton and Chorley also met the canal packets.

In Ireland, the Grand Canal was well known for its passenger services from Dublin to the Shannon, and for the hotels that

LANCASTER CANAL.

THE SWIFT

PACKET BOATS

SAIL DAILY,

TWICE in each direction between PRESTON and KENDAL, and THREE TIMES in each direction between LANCASTER and PRESTON,

IN CONNEXION WITH THE RAILWAY TRAINS.

SOUTHWARDS.

From	H.	M.	H.	M.	MILES	H.	M.
KENDAL			6	30	0	8	30
Crooklands			7	30	9	9	30
Holme			7	45	11	9	45
Tewitfield			8	30	14	10	30
Carnforth			8	55	19	10	55
Bolton			9	5	21	11	5
Hest Bank			9	20	23	11	20
LANCASTER	6	0	9	50	27	11	50
Galgate	6	20	10	20	32	12	20
Clifton Hill	6	45	10	45	35	12	45
Garstang	7	20	11	20	40	1	20
Roebeck	7	50	11	50	45	1	50
Swilbrook	8	15	12	15	50	2	15
Salwick	8	25	12	25	52	2	25
Arrive at **Preston**	9	20	1	20	57	3	20

NORTHWARDS.

From	H.	M.	MILES	H.	M.	H.	M.
PRESTON	11	0	0	1	30	5	0
Salwick	11	35	5	2	5	5	35
Swilbrook	11	45	7	2	15	5	45
Roebeck	12	10	12	2	40	6	10
Garstang	12	40	17	3	10	6	40
Clifton Hill	1	15	22	3	45	7	15
Galgate	1	40	25	4	10	7	40
LANCASTER	2	20	30	4	50	8	20
Hest Bank	3	0	34	5	30		
Bolton	3	15	36	5	45		
Carnforth	3	25	38	5	55		
Tewitfield	3	50	43	6	20		
Holme	4	40	46	7	10		
Crooklands	4	55	48	7	25		
Arrive at **Kendal**	6	20	57	8	50		

FARES.

	First Cabin.	Second Cabin.
Between KENDAL and LANCASTER, or LANCASTER and PRESTON	3s.	2s.

THE RAILWAY TRAINS

From PRESTON	9 45 morning
— Do.	2 20 afternoon
— Do.	4 20 afternoon
From LIVERPOOL	8 45 morning
— Do.	11 0 do.
— Do.	2 30 afternoon
— MANCHESTER	9 0 morning
— Do.	11 15 do.
— Do.	2 45 afternoon

FORM DIRECT CONVEYANCES WITH THE PACKET BOATS.

The Railway Trains at 9 45 and 2 20 are in immediate connexion with the GRAND JUNCTION TRAINS from Manchester.

Omnibuses between the Packets and the Railway.

Coaches to and from Bolton and Chorley, meet the Packets at the Canal Wharf, Preston.

Parcels carefully conveyed and delivered free of Portorage.—It is particularly requested that all Parcels for any place between Preston and Kendal, sent by the Railway Trains, should be marked "To be be forwarded by Canal Packets."—Small Parcels between Preston and Lancaster, or Lancaster and Kendal, charged only SIXPENCE each, other distances equally moderate.

CANAL OFFICE, LANCASTER, JUNE, 1839.

OAKEHALL, PRINTER, PRESTON.

14 The Lancaster Canal swift boats connect with trains to Liverpool and Manchester, and coaches to Bolton and Chorley

the company built along the way. There must have been considerable romance on these boats, on a fine evening after dinner, watching the countryside moving by, the postilions bobbing up and down on the horses, the captain in his blue frock coat, scarlet waistcoat and shiny hat. Yet most travellers did not see it like this: Martin Kelly, for instance, on the Ballinasloe canal boat, as Anthony Trollope describes him:

> Reading is out of the question. I have tried it myself, and seen others try it, but in vain. The sense of motion, almost imperceptible, but still perceptible; the noises above you, the smells around you; the diversified crowd, of which you are a part; at one moment the heat which this crowd creates; at the next, the draught which a window just opened behind your ears lets in on you; the fumes of punch; the snores of the man under the table; the noisy anger of his neighbour, who reviles the attendant sylph . . . the loud complaints of the old lady near the door, who cannot obtain the gratuitous kindness of a glass of water; and the baby-soothing lullabies of the young one, who is suckling her infant under your elbow. These things alike prevent one from reading, sleeping, or thinking. All one can do is to wait till the long night gradually wears itself away, and reflect that, 'Time and the hour run through the longest day.'

In Ireland, too, swift boats were put on to supplement the slower and cheaper services, until the railways ended them both.

On the Continent, passenger carrying on the canals was commonest in Belgium and the Netherlands. As a young man, the artist Sidney Cooper went there in 1830, and wrote:

> The means of transit from Rotterdam to The Hague was by the canal, on board a conveyance called a 'Traekschout' –

a sort of boat-omnibus, towed along by two horses – a very easy and pleasant style of travelling. The fore part of this boat was set apart for the poor people, and the cabin and stern for gentlemen, who were provided with pipes, tobacco and cigars, a wooden shoe, or sabot, being placed on a small table for a spittoon.

Earlier, in 1815, Robert Southey had travelled from Bruges to Ghent:

After breakfast we embarked in a Trekschuit, which has obtained the reputation of being both the best and the cheapest public conveyance in the world. The scene at the point of embarkation, by the iron gates at the end of the canal, was delightful for anyone who has a painter's eye. Vast numbers of people were arriving, many in carriages of sundry odd forms . . . an English chariot which we took on board was some nuisance by the room which it occupied . . . The Trekschuit has a canopy at the stern, somewhat of a bell-shape . . . on the top of this is a painted plume of feathers. There are two cabins below, and between them, kitchens, *commodities* (the word is a commodious one), and heaven knows what beside. It was full of passengers, of whom a great proportion were English . . . when we met another vessel, or wanted to have a bridge wheeled round for our passage, the man at the helm either blew a pocket whistle or rang a bell, or set up a Flemish halloo . . . We had an excellent dinner, included in the fare of five franks a head . . . When we reached the quay, it was crowded with spectators. Some thousands certainly had assembled, as if all the idle part of the population regarded the arrival of the Trekschuit as a sight, and were waiting for it.

Boys and porters contended for the passengers' luggage, and coaches were waiting to take the Southeys to their hotel.

The author of a little book published in 1834, *What may be*

Done in Two Months, described the passage boat from Ostend to Bruges:

> The *Elégante Messagère* deserved her name; for never was a cleaner or better fitted-up packet-boat: the little cabin, with its beautiful sofas, mirrors and chairs, rivalled the nicest drawing-room: and breakfast was served in it with much propriety and order . . . The canal is broad and was nearly full; so that we were above the banks: the boat was dragged by two horses; and we proceeded along at the rate of four miles an hour. We reached Bruges in about three hours; and on paying our fares, were agreeably surprised to find it only two francs each person including breakfast: the distance about thirteen miles. We were landed about a mile from the town, and found a diligence waiting the arrival of the barge, to carry the passengers to the Ghent packet.

A remarkable service was provided for a time on the 150-mile-long Canal du Midi in southern France. Mail-carrying packet boats connected at Sète with ships to Mediterranean ports, and at Toulouse with other boats to Bordeaux. To begin with, to save time, boats did not pass through flights of locks, but passengers and mail were transferred over twenty times from one boat to another up or down the locks. This system must have given everyone a lot of exercise and not a little cause for patience before it was replaced by slightly slower craft which did pass through all locks except the staircase at Béziers.

River travel was, of course, extensive. The day boats that now work over the popular section of the Rhine between Cologne and Mainz had their counterparts then: in 1833, the steamer *Stadt Mainz*, a 'beautiful vessel', in which the writer was

> most comfortably accommodated, both as to space on deck, and elbow-room in the grand salon

did the same trip, taking two days, with an overnight stop at Koblenz. From Mainz it continued up river to Karlsruhe, and in the other direction down to Rotterdam, with overnight stops during which passengers were accommodated ashore.

All this was small beer to the North American passenger boats, built for journeys that might take several days on the canals leading westwards to the Lakes or the Ohio, and run by packet companies, perhaps the Pioneer or the Express. They were mostly of similar design, about 70 ft long and 14 ft wide, with a crew room in the bow, then the ladies' dressing-room, curtained off from the main cabin, 36 to 45 ft long, which was men's sleeping-quarters at night and saloon by day. Behind this was the bar, and at the stern the kitchen, probably with a Negro cook who was also the bartender. Some larger boats also had a freight compartment.

Mrs Frances Trollope, Anthony's mother, travelling from Schenectady to Utica on the Erie Canal, was not favourable:

> The accommodation being greatly restricted, every body, from the moment of entering the boat, acts upon a system of unshrinking egotism. The library of a dozen books, the back-gammon board, the tiny berths, the shady side of the cabin, are all jostled for in a manner to make one greatly envy the power of the snail; at the moment I would willingly have given up some of my human dignity for the privilege of creeping into a shell of my own . . . In such trying moments as that of *fixing* themselves on board a packet-boat, the men are prompt, determined, and will compromise any body's convenience, except their own. The women are doggedly steadfast in their will, and, till matters are settled, look like hedgehogs, with every quill raised, and firmly set, as if to forbid the approach of any one who might wish to rule them down.
>
> With a very delightful party of one's own choosing, fine temperate weather, and a strong breeze to chase the mosquitoes, this mode of travelling might be very agreeable, but

> I can hardly imagine any motive of convenience powerful enough to induce me again to imprison myself in a canal boat under ordinary circumstances.

But it had been very hot:

> Pretty well fagged by the run by day, and a crowded cabin by night; lemon-juice and iced-water (without sugar) kept us alive. But for this delightful recipe, feather fans, and eau de Cologne, I think we should have failed altogether; the thermometer stood at 90°.

Earlier, in a day boat on the short Chesapeake & Delaware Canal, she had been more favourable to canal travel:

> we got on board a pretty little decked boat, sheltered by a neat awning, and drawn by four horses.

Charles Dickens' account of the canal boat on the Pennsylvania Canal main line from Harrisburg to the Alleghenies is the best known to Englishmen: the men sitting at little tables down each side of the main cabin, the ladies' room partitioned off behind a red curtain, a boy riding the leader of the three towing horses, the creaking rudder. On the small deck, made smaller by the passengers' luggage stacked under a tarpaulin, one had to

> duck nimbly every five minutes whenever the man at the helm cried 'Bridge!' and sometimes, when the cry was 'Low Bridge', to lie down nearly flat.

At night the ladies slept behind their curtain, the 28 men in 'three long tiers of hanging bookshelves, designed apparently for volumes of the small octavo size', suspended on either side of the cabin, each with its 'microscopic sheet and blanket'. The night spent with all the windows shut, amid fear of the collapse of the heavy gentleman's bunk above one, a chorus of snores, a storm of spitting and the sound of the steersman's bugle each time the boat approached a lock. The morning wash between 5 and 6 AM,

> standing on the deck fishing the dirty water out of the canal with a tin ladle chained to the boat by a long chain; pouring the same into a tin basin (also chained up in like manner); and scrubbing my face with a jack-towel;

breakfast when the tables were put together as one, a waiter afterwards shaving the gentlemen; the smells of food, drink and sleeping people. And yet

> there was much in this mode of travelling which I heartily enjoyed at the time, and look back upon with great pleasure.

Getting out of the cabin to the fresh air and the cold water; the brisk walk along the towing path; the experience of the early morning:

> the lazy motion of the boat, when one lay idly on the deck, looking through, rather than at, the deep blue sky; the gliding on at night, so noiselessly, past frowning hills . . . the shining out of the bright stars undisturbed by noise of wheels or steam, or any other sound than the limpid rippling of the water as the boat went on; all these were pure delights.

On other boats sailing on other waters, we too have known what he knew, and have experienced delight.

The railways came, and most of the packet boats were taken off soon afterwards, though some lasted for twenty or more years. But what had begun as a necessary public service was transformed into holidaying, and many canals saw the coming of boats built specially for enjoyment: excursion steamers, like the *Countess of Ellesmere*, running as early as 1838 on the

Mersey & Irwell from Manchester to Pomona Gardens, where Salford docks now are, and the later *President*, *Punch* and *Judy*; the *Queens* of the Forth & Clyde, who ruled from 1893 to 1940; or MacBrayne's vessels through the Crinan and Caledonian canals. These were the forerunners of the Continental tripboats of today.

In Victorian times, a Sunday school or village outing was often made in a hired barge. As in 1840,

> That Mr. Hopkins have free permission to take the children from the Newton Heath Sunday School to Chadderton Park by canal as requested.

Many an old person living near the waterways still remembers such an excursion as a highlight in the child's round of the country year. This activity, too, has become transmuted into the society or club outing of today, in a narrow boat or barge fitted with seats and serving refreshments.

In these times, too, came the beginnings of pleasure cruising, each venture so exceptional, so truly exploratory, that a book about it almost always followed. On the Continent, the intrepid three who in the 1850s rowed from Rouen across the rivers and canals of France to Mulhouse, and then down the Rhine to Cologne, were like this; or Robert Louis Stevenson and his friend, damply canoeing from the Schelde by way of Brussels to the Sambre et Oise Canal and Pontoise in the 1870s.

In Britain, it began in the 1860s: A *Trip through the Caledonian Canal* in 1861; *The Thames to the Solent by Canal and Sea*, 1868; *The Waterway to London . . . in the Voyage on the Mersey*, *Perry*, *Severn and Thames*, *and several Canals*, 1869; *A Canoe Cruise from Leicestershire to Greenhithe*, 1873. Then a pause, followed by the long series of *Oarsman's Guides* and similar publications for those who rowed or canoed on rivers and canals, and early cruising books such as C. J. Aubertin's *A Caravan Afloat*. And so to the first of the motor-boating handbooks, George Westall's *Inland Cruising on the Rivers and*

CRINAN CANAL

NOTICE.

CHILDREN and Others are hereby Prohibited from Running along the Canal Banks after the Passenger Steamer; and Passengers are requested not to encourage them by throwing Money on to the bank.

Children are further warned not to throw Flowers into the Boat.

SALE OF MILK.

THE SALE OF MILK on the Property of the Canal Commissioners is only permitted on the understanding that no annoyance is caused to Passengers.

Any person who, by urging to purchase, or otherwise, inconveniences or annoys any Passenger will be prohibited from selling, and, if necessary, dealt with according to law.

Passengers are requested to report any such case to the Purser, and also to point out the delinquent to the nearest Lockkeeper or Canal Official.

L. JOHN GROVES,
SUPERINTENDENT.

Crinan Canal Office.
Ardrishaig, 30th June, 1887.

15 A Crinan Canal notice of 1887

Canals of England and Wales in 1908. But canal cruising still remained very exceptional until after L. T. C. Rolt's *Narrow Boat* of 1944. A great change has taken place since then.

British canals nowadays see many types of pleasure boat: a few hotel and hostel boats, offering a week or a fortnight of leisurely travel; day boats, running excursion trips; and a rising tide of narrow boats converted to floating homes, motor cruisers, small outboards, and canoes. The craft change, but the life of the canals goes on.

CHAPTER 8

How Canal Companies Worked

We shall wrong history if we think of canal companies with the same mental picture we have when we use the word 'company' today – impressive offices, keen-eyed directors, busy administrations, and all. They were small – few employed two hundred people – and by the nature of the business those few were scattered along a curving line of water. A toll-taking roll did not demand initiative, and therefore direction from the centre was usually weak; but engineers, toll-collectors, lengthsmen (maintenance men) and lock-keepers needed above all to be reliable, to be trusted to work honestly and conscientiously with little supervision. A race of canal men of this kind grew up. It became for many a hereditary occupation, and today men with the same qualities still do similar work, some of them the descendants of those who learned their trade in the canal age.

As I write, past issues of the British Waterways Board's monthly magazine *Waterways* lie on my desk. Mr Samuel Price has retired: he was born at the Canal House at Sneyd junction on the Birmingham Canal Navigations, where his father was a district inspector, as was his uncle. He joined the BCN as a boy, and spent his life in the canal service. A little earlier it was Harry Degge, clerk in the Birmingham area engineer's office, after 52 years on the Birmingham Canal Navigations, for whom his father and his grandfather also worked. Later, Richard Turner died, aged 87, a retired foreman-carpenter on the Leeds & Liverpool Canal. His father, grandfather and great-grandfather were all Leeds & Liverpool men, and he himself lived all his life at the lock-house at

Banknewton. J. C. Bradley, now manager of Sharpness dock, is one of the seventh generation of canal and river navigation men. Many an engineer, clerk, committeeman or chairman was also succeeded by his son. Such was the life and atmosphere of the canal business.

At the centre was the shareholders' meeting, perhaps two or three dozen men, perhaps only half a dozen, some holding proxies for non-attenders. Some companies paid an expense allowance, 3*s* 6*d* a time from the Droitwich in 1786, or refunded expenses incurred. Some gave a free dinner, like that the Swansea Canal company provided at a cost of 5*s* per head, 'beer and a genteel dessert to be included', at which constant traders were asked to dine with the shareholders, a sensible piece of public relations. But free drinks could prove fatal, as the old Union Company found when they sadly reported in 1796 that the great expense of shareholders' meetings was

> in great measure owing to diverse persons, who have no interest in the Concern, drinking and taking away considerable Quantities of Wine and Liquor.

It was to be food only in future.

This meeting elected a committee, usually from seven to fifteen, of whom up to half a dozen were likely to be regular attenders. It could meet monthly on a big canal, three or four times a year on a small one, and took most of the decisions necessary except toll alterations and decisions on dividends. Committeemen were not paid during the canal age, but usually got an expense allowance and a free meal before their horses took them home. Meetings took place in the canal offices or a local inn, and in the summer some part of the canal would be inspected, on the committee's own boat or one borrowed from a neighbouring company. Meetings held on board would then be date-lined, perhaps from the 'elegant and commodious' *Savile* yacht, which belonged to the Calder & Hebble. The

Derby Canal Company arranged that such perambulations should be publicized by handbill, so that anyone with a grievance could meet the committee on the spot.

The senior officers were a clerk, treasurer, engineer, accountant and agent or manager; whether these were part or full-time, and whether one man did more than one job, depended on the size of the company.

Clerks kept the minutes, and executed the orders of committee and shareholders' meetings. They were usually local solicitors who took on canal work for an annual fee. In small concerns they also acted as agents.

Accountants had to collect toll-takings regularly from the toll-keepers and check their books, and also from those carriers who were given a credit account, hand the receipts to the treasurer, keep the company's books, and produce annual figures.

Treasurers in the early days were private individuals of substantial fortune, often industrialists. Seldom paid, and usually asked to give a substantial security bond – Abiathar Hawkes, glassmaker, treasurer of the Dudley Canal, put up a £5,000 bond and sureties for another £5,000 in 1793 – it was expected that they would recompense themselves by using the company's balances in their own businesses. But, as banking spread widely from about 1790, so did canal companies from the early decades of the 1800s begin to appoint banking concerns as treasurers, though sometimes in the name of the bank's partners rather than the business itself. Wise companies, however, kept accounts at different banks to minimize the risk of failure as well as for the more convenient payment of dividends: the Oxford Canal had four in 1790, its main account with Thos Walker & Co at Oxford, others with Child & Co (London), Bignell & Co (Banbury) and Little & Co (Coventry).

When agents or superintendents were appointed – we should now call them general managers – they often provided the company with a drive that no one else but the committee chairman was placed to give. Thomas Brewin was appointed full-time agent of the Dudley Canal in 1812 at £250 per annum and expenses, having sat on its committee for the three previous years. The following year he was back for a seven-year contract which included a commission on extra tolls earned, which he swapped in 1819 for a £50 rise and permission also to run a wholesale trading business in coal from mines on the Dudley Canal to wharves on those connecting with it. He bought his own company's (and others') shares, when they were low, chased the committee to make improvements, got himself a seat on the Stratford-upon-Avon Canal board to represent his company's interest, and saw his shares rise steadily; he bought more and some for his wife. In 1839, when his salary was raised to £400 per annum, the committee minuted their company's appreciation of his

> management . . . to which its present welfare & prosperity is in a great degree attributable.

When he retired about 1846, his company had just amalgamated with the Birmingham Canal Navigations, who issued their shares in return for those of the Dudley; so his financial position was snug.

The engineer who had built a canal usually moved on, to be replaced by another who had the different task of maintaining and improving it. On most canals he was a professional, perhaps a man retired from the business of building and contracting for canals, like Thomas Sheasby, who had done contracting for the Cromford and Coventry companies, had built the Swansea and worked on other canals. In 1801, the year after it was opened, he was taken on by the Warwick & Birmingham company at 100 guineas a year, and 20 guineas for his horse. He held the job for four years, and then retired through ill health. When Arthur Gilbody was appointed engineer of the Mersey & Irwell (which was both a navigation company and a carrying business with sailing flats), his pay being £200 per

TOWNSEND.H

George Inn

HUDDERSFIELD

Apr 27		L. S. D
	Mr Leer Wine	0.3.9
	Tea	0.1.0
	Suppers	0.6.0
	punch	0.4.0
28	Tea	0.4.0
	Dinner	0.16.0
	Ale	0.2.0
	port	0.10.6
	Sherry	0.5.0
	Tea	0.8.0
	Hay Corn	0.7.6
	paper	0.0.2
	punch	0.0.6
		4.11.5
	Mr Buckley Dinner Wine	2.3
		4.13.8
	Servants Horses &c.	1.6
		4.15.2

Huddersfield April 29th 1802

Received of the Huddersfield Canal Company by John Rooth – four Pound fifteen Shillings & Twopence, for the Expense of Sub Committee when Examining accompts as per annexed acct.

Jn Townsend

£4..15.2

16 (*a*) and (*b*). The accounts sub-committee of the Huddersfield Canal met on April 27th, 1802, at the George Inn, Huddersfield. They had tea and dinner with drinks while their servants and horses were also provided for. The innkeeper's bill was paid by the canal company's agent, John Rooth

annum, a house, and keep for a horse and cow, his duties were to superintend the whole line, see to the repair of weirs, locks, bridges, sailing flats and tackle, see that voyages were not interrupted, buy timber, and oversee the company's farms and horsekeepers.

Or he could be one who had learned his business from the construction engineer. Christopher Staveley started in 1790 as a surveyor, working with the great William Jessop of the Leicester Navigation. Having learned engineering thus, two years later he became resident in charge of direct labour construction. This finished, he was given a part-time contract as engineer which was periodically renewed until he died in 1827, during which time he worked also for other companies. He was succeeded by his son Edward, who two years before had been joined with him as joint engineer. But, sad to say, Edward was not the man his father was. He worked for six years, then cashed a couple of quick cheques on the company's account, briskly left the country with £1,400 that was not his and was never heard of again.

Or a man could combine the jobs of agent and engineer. John Warner joined the Coventry company in 1795 as agent, but learned engineering. By 1815 he was earning 300 guineas plus a horse. He died in 1820, and was succeeded by a professional engineer, who then had to learn the business of an agent. John Sinclair had been one of Thomas Telford's men on the Highland roads and the Holyhead road. He came at £200 and a house, but quickly showed his ability, for by 1832 he was up to £400, with permission to do outside professional work when he could fit it in.

Distinctions of office were less clear then than one would now expect. Take Richard Tawney of Oxford, who in 1794 was appointed agent and engineer of the Oxford Canal. The Tawneys were well-known local people, Alderman Edward Tawney being on the canal committee in 1792 and 1794. Tawney began at £150 per annum plus expenses, and by 1810 had risen to £300 per annum and £150 expenses of rent, horse-hire and so on, to which annual presents of 50 guineas were often added from 1813 onwards. However, he must also have had considerable private means, for in 1819, with his brother Charles, an Oxford brewer and later also on the canal committee, he bought the Banbury New Bank, mainly as a career for his second son Henry, re-selling it in 1823 to the Gilletts on condition that Henry entered it. Here was a versatile man – financier, manager, engineer, with relatives on the committee: but a man of the eighteenth rather than the nineteenth century, when professionalism was more sought after. So when Tawney died in 1835, a professional, Frederick Wood, took over on a part-time basis. He had helped to survey and carry out the straightening of the canal's northern section, just finished, and worked also for other companies. In 1840, he was appointed full-time at £500 per annum, but, alas! being a professional, he left in 1853 for the London & North Western Railway, though agreeing also to look after the canal part-time for £250.

Beneath the clerk and accountant were the office staff and toll-collectors: the agent controlled the lock-keepers, wharfingers, and carrying staff if the canal ran its own boats; the engineer, the masons, carpenters and other craftsmen, the lengthsmen and labourers who worked along the banks; and the reservoir-keepers.

There were two to six toll-collectors on an average canal, employed to check cargoes, gauge boats, and receive the tolls. They were picked men who had given security for their honesty. One at Lenton on the Nottingham Canal in 1806 got £50, which in 1829 had risen to £80, and in 1835 to £90, but in 1840 became £70 and a house. Because their maintained honesty was so important, they had to be kept free from situations where pressure could be put on them to be dishonest – hence the same company told them they must not trade with the boatmen, or sell them ropes. But if they liked to get up in the night to pass boats after hours (the canal was open from one hour before sunrise to one hour after sunset), there was no objection to their taking a shilling tip. Some gave way to temptation, like the man on the Coventry Canal in 1843, who was caught borrowing money from boatmen, and was then found to have embezzled some of the tolls.

John Brown, the Oxford's toll-collector at Hawkesbury Bar, junction with the Coventry Canal, in 1790 got 18*s* a week, but had to find his own firearms (the clerically-controlled Oxford Canal was definitely the church militant); he was given an assistant at 10*s* per week, and 2*s* for lodging, so John could stay in bed at night. On the Trent, with rapidly increasing business, toll-collectors who in 1784 were getting £20 and £30 per annum had risen by 1792 to £40–£60, by 1802 to £80, and by 1813 to £120 per annum. But, as on the Oxford, the job was not without risk, for in 1813 they were

> provided with a Blunderbuss and Bayonet at the Expense of the Company,

in case thieves should break in and steal the tolls.

Nowadays, lock-keepers are few and far between, but in the heyday of the canals almost every lock had its lock-house. The keeper's job was not primarily to work the locks – the boatmen did that, though in competitive days they were expected to help in the interests of speed. They were there to make sure the boatmen did not damage the lock gates or structure, and observed the bye-laws, to ensure that water, scarce on most canals and worth two or three shillings a lockful, was not wasted, and to prevent boatmen quarrelling for precedence at the lock.

Lock-keepers had many temptations. They could take bribes for giving preference to one boat over another, either in cash or by raking out some coal from the canal after a boatman had accidentally let it fall in; they could go in for selling drinks, or groceries, or ropes; or have a little business of their own, like a smallholding. All these were forbidden over and over again, so often that surreptitious disobedience must have been pretty common. Minor relaxations were often allowed – for instance, the lock-keeper at the river lock at Lydney was allowed to do shoemaking, presumably between tides. Depending on the management, lock-keepers and other employees had to watch their step during such off-duty times as they had. Two on the Monmouthshire were discharged in 1801 for taking part in riots at Pontypool. On the other hand, those on the Oxford, with a clerical majority on the board, probably took care to be seen occupying the pews the company rented at churches along their line. Therefore lock-keepers were fairly well paid by contemporary standards – on the Stourbridge Canal in 1828 they got 15*s* to 16*s* and their houses and gardens; this rate was raised to 18*s* to £1 in 1834 because of night work, and in 1847 reached 30*s*. On a rather less busy canal, the Grantham, in 1839, the pay was 18*s* and a house. Sometimes free coal was given, to prevent it being stolen, or a coal allowance paid. There might be a candle allowance for night work. Sometimes there would be a reward for informing against boatmen when bye-laws were broken. Nearly always a pension was paid for long service – the Glamorganshire was paying 5*s* a week in 1837, the Staffs & Worcs 6*s* in 1809, the Warwick & Birmingham, exceptionally, 8*s* a week, or half wages, in 1830.

Wharfingers were men in charge of a public wharf, where goods were loaded and unloaded, and stored either in the open, like coal, or under cover in warehouses. They might also be local agents, working to the manager. Theirs was a responsible job, and one open to temptation of theft, or of collusion in theft by others. Therefore such men were reasonably well rewarded. In 1800 the Stourbridge paid 50 guineas a year for a town wharf, in 1795 the Oxford £52 for a similar one, and 12*s* for two country ones – with a house in all cases, of course. They soon had to raise their rates as business increased; by 1843 the town rate had doubled to £100 per annum, tapering off to £80 and £60 at smaller wharves. A really big wharf such as Manchester on the Rochdale Canal earned £400 per annum for its wharfinger. On the other hand, as competition grew, there were snags to the job:

> in case of need, the Wharfinger shall stop up at Night to render every facility for the despatch of all vessels which are under the necessity of sailing during the night.

Sometimes wharfingers got a percentage on wharfage charges as well as their pay, as at Stourport.

The man in charge of a big warehouse, immediately under the wharfinger, got £52 at Grantham in 1815. Warehouse porters got less, but there was

> the customary allowance for Ale of 1½*d* for every Ton of Grain in bulk, weighed or measured and landed into the Warehouses.

Skilled maintenance craftsmen presumably got about the same as their opposite numbers outside the canal business – such as the master-carpenter on the Glamorganshire Canal in 1825 who was paid £6 a month and another 10*s* for horse-hire, as well as his house. Later his rate came down to £5, but was back at £6 in 1846. Ordinary masons and carpenters on that canal were getting 21*s* a week in 1848. On the Derby Canal the carpenters' rate in 1812 was 24*s* with a house, or 28*s* without one. In 1841 on the Staffordshire & Worcestershire a carpenter was getting 2*s* 10*d* a day, a sawyer 3*s* and a stone-mason 24*s* a week.

Quite an important skilled man was the molecatcher, who prevented the little gentleman in black velvet from undermining embankments. He got £18 per annum part-time on the Brecknock & Abergavenny in 1804 for looking after 30 miles of canal. The full-time man on the neighbouring Glamorganshire got 15*s* per week in 1810.

Unskilled labourers on the Derby Canal got 13*s* and a house in 1839. Houses were important – in 1840 the Neath company built six especially in order to attract good labourers to the service of the canal. Much the same rate – 12*s* to 14*s* – was paid on the Swansea Canal. One extra they usually got was free beer for especially hard or unpleasant work.

Lastly, there were the office workers – the trained men well enough paid, like the Glamorganshire's book-keeper at the big interchange wharf at Navigation House, or the check clerks at Cardiff and Merthyr, each paid £100 per annum in 1825. But the minor clerks were poorly paid as such usually were, like the check clerk on the Erewash Canal in 1786 who petitioned that his £20 a year was not enough for 'the decent Maintenance of his Family', and was given a rise to £26 – surely all the little Cratchits must have rejoiced that evening. His opposite number at Harby on the Grantham Canal in 1804 got about the same, 8*s*, but also a house, but he at Redhill on the Loughborough Navigation moved *up* to £20 in 1793.

Nearly all these jobs were pensionable for long-service men, at a rate usually of about 5*s* a week, but as high as £1 for a senior toll-collector, or 10*s* for a master-craftsman, and as low as 2*s* for a labourer. Others got gratuities.

Hours of work were hardly defined – they were what the job required. If a lock-keeper wanted time off, then his wife or son took his place. If a bank burst, men worked till it was repaired, and were probably given a rest afterwards. Officially, hours were about twelve a day for six days a week on the Staffordshire & Worcestershire in 1785, and much the same a hundred years later. Occasionally, piece-work rates were substituted for day rates, especially for labourers, if the job could be measured.

Women often helped their husbands, for instance at lock-keeping, but were seldom independently employed, except on the Weaver Navigation where this often happened. The Monmouthshire company allowed a widow to succeed her husband as agent or toll-collector at the same wages in 1827, and the Stourbridge had a woman lock-keeper in 1830. In 1886 the Derby company allowed a widow to take on her husband's toll-collecting job, but three years later gave her notice, minuting that they considered the duties could not efficiently be discharged by a woman. Widows were often given a gratuity of perhaps £5 on the death of their husbands and a period of grace to leave their houses; occasionally a small pension – in 1835 on the Stourbridge Canal, £5 a year. The Leicester Navigation company on one occasion bought a cow for £10, to give to the widow of a labourer who had been killed at work, so that she would have a means of livelihood.

Union activity was almost non-existent in the canal business until railway times. So wage rates varied to some extent on a personal basis, such as years of service, hardship, hours of work, or responsibility, but more often on a group basis according to how the company was doing. For instance, in 1817 the Derby Canal reduced all carpenters' wages by 3*s* and labourers on their tramroad by 1*s* a week, and in 1829 again cut their labourers – presumably they were judging their own rates against those prevailing outside. On canals such as those where the canal owners were also ironmasters, these were concerned to see that the two rates were kept in line, so that ironworkers did not leave to work on the canal or vice versa. There was a general reduction on the Glamorganshire in 1822, a general rise in 1825, and another in 1836, none of them connected with the finances of the canal company itself.

Just occasionally the staff got gratuities, such as the long-service lock-keepers on the Stourbridge in 1846, or a bonus, like the clerks on the Staffordshire & Worcestershire who got leave and a sum of money to enable them to see the Great Exhibition, or a free celebration, when all the workmen of the Weaver got a 5*s* dinner with drink at Queen Victoria's coronation. Sometimes, too, the company operated a simple sort of health insurance and disablement compensation scheme, encouraged the men to form a sick club, or contributed to hospitals where their staff could go.

It is a picture of comparatively small-scale business. One, also, of independence. For, until railways came, the vision of

CANAL PROPRIETORS: (*above*) the Earl of Ellesmere had his private barge and postilions on the Bridgewater Canal; (*below*) the North Staffordshire Railway directors' launch *Dolly Varden* on the Trent & Mersey Canal

BOAT PEOPLE: (*above*) a boat family and pair of narrow boats at Braunston top lock on the Grand Junction Canal about 1912; (*below*) young people on the Regent's Canal in 1902

CANAL WORKERS: (*above*) the lock-keeper; (*below*) the barge-horse

CANALS FOR PLEASURE: (*above*) *Gipsy Queen* on the Forth & Clyde Canal in 1937; (*below*) *Linnet* on the Crinan Canal about 1900

Calder and Hebble Navigation Office.

HALIFAX, March 8th, 1816.

ROBBERY.

FIFTY Pounds Reward.

Whereas

The WAREHOUSE belonging the Company of Proprietors of the Calder and Hebble Navigation, situate at Cooper-Bridge Wharf, near Mirfield, was broken open and feloniously entered by some Villain or Villains in the Night betwixt the 5th and 6th Days of March Instant, and the following Pieces of Cloth taken out of a Bale, deposited in the said Warehouse, marked B in a Diamond, Viz,

1	Piece Drab,	marked	No. 1797,	measuring	36	Yards,
1	Do. Blue	Do.	1817,	Do.	36¾	Do.
1	Do. Do.	Do.	1820,	Do.	36¾	Do.
1	Do. Do.	Do.	1818,	Do.	37¾	Do.
1	Do. Olive	Do.	1742.	Do.	39	Do.
1	Do. Do.	Do.	1739,	Do.	40½	Do.
1	Do. Do.	Do.	1743,	Do.	40¾	Do.
1	Do. Do.	Do.	1744,	Do.	41	Do.

A Reward of FIFTY POUNDS

Is hereby offered to any Person or Persons who will give such Information as shall lead to the Conviction of the Offenders; and if any of the Perpetrators or Accomplices in the said Robbery, will inform of the other or others of them, he shall in like manner be entitled to the said Reward, and every Means used to obtain his Majesty's Pardon,

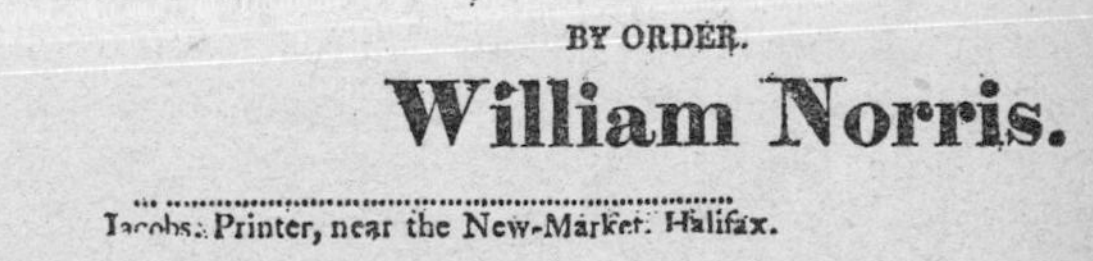

BY ORDER.

William Norris.

Jacobs, Printer, near the New-Market, Halifax.

17 The Calder & Hebble take action against thieves

those committeemen who worked towards creating an efficient countrywide system of waterways was usually overcome by the concentration of the locals upon the preservation of their own company's position.

There were times when companies came together: in 1769, for instance, when five agreed on common standards for narrow boat locks; in 1797 when William Pitt's proposed tax on goods carried by inland navigation caused a scurry of joint meetings, and in the same year when on the initiative of the Trent Navigation, about ten concerns in the East Midlands joined to operate a common system of gauging boats; on more than one occasion over the rating of canals by local authorities, and in 1844 when bills to authorize companies to vary their tolls on different parts of the line, and to become carriers, were being prepared in London. After that, with a common enemy, communications became more usual, and led to the setting up of the Canal Association.

Otherwise, each company was wary of its neighbours. Communication was slow, for committees seldom met even as often as monthly, and a letter approved by one might wait two months or more for an answer by another. If the subject was a trading one, a proposed change of toll or of opening hours, slow action might follow. But if a concession were asked for, then negotiations might be short, ending in refusal, or long, and result in some sort of adjustment.

In those days, the income from a canal share was considered to confer a property right analogous to that in land. Therefore, if income were threatened, perhaps by a new canal which might change the traffic pattern and affect the company's dividend, that was thought to require compensation. Hence the compensation tolls of so many pence or shillings a ton, that new companies had to pay for being allowed to join an existing canal, tolls which often distorted trade patterns over a generation, as did that of 10½*d* a ton, where the Ellesmere & Chester (later the Shropshire Union) joined the Trent & Mersey at Wardle near Middlewich. Hence also the flat refusals to allow a junction, as that the Trent & Mersey had fifty years earlier given to the Chester Canal company when they wished to make a connexion at the same place. Worcester Bar at Birmingham was the best known result of such a refusal, a physical bar between two canals over which goods had to be transhipped by crane for many years.

The Grand Panjandrum, in the forty years to his death in 1803, was the Duke of Bridgewater himself, owning his prosperous waterway from Worsley to Manchester and on to Runcorn, his coal mines and his property in and around Manchester. He had no wish to encourage canals anywhere, unless they benefited him, and competitors were far from welcome. When the Rochdale Canal, promoted in 1792, asked to join his waterway he

> positively objected to a junction between it and his Navigation at Corn-brook, or at any other place,

but later agreed to accept a compensation toll of no less than 3*s* 8*d* a ton for allowing a junction, such toll 'to be invariable for ever', and so awesome was the Grand Panjandrum that the Rochdale promoters minuted that the

> Proposals of his Grace are reasonable and ought to be accepted,

and added that their Chairman should

> take the first Opportunity of presenting the grateful Thanks of this Committee to his Grace for his good intentions.

Fortunately, by the time of the Rochdale Act, he had seen his own interest in encouraging through traffic from Yorkshire over his own canal, and brought down his compensation toll to 1*s* 2*d*. It was still quite enough. When the Mersey & Irwell were trying to incorporate their concern and modernize its company structure, the Duke opposed the bill with his great Parliamentary influence, and got it withdrawn. They then asked him if he would oppose another one; he said he would, and if they altered it to meet his criticisms, he still would. In the end he relented just enough to let the bare minimum of reform go through. In 1803 the great, difficult man died, but his spirit lived on in the Trustees who thereafter controlled his canal, and in their manager, Robert Haldane Bradshaw, as awkward and energetic a character as the canals ever produced. Samuel Smiles has the story that when the Trust was offered participation in the Liverpool & Manchester Railway, his reply was, 'All or none', meaning that if the Trust could not control the railway they would fight it. Samuel Smiles and others have laughed at this reply as one given by a last ditcher, incapable of foreseeing the future. But, fifteen years after the railway was opened, the waterways were still carrying three-quarters of a greater total traffic between Liverpool and Manchester, and when today we watch the world's goods brought to Manchester up the Ship Canal in many times the quantity that they are carried by rail from Liverpool, do we say the old man was wrong to back water against rail?

The result of canal company organization was, on the whole, quite good local transport service, but poor facilities for long-distance trade. Improvements of course there were, but piecemeal and painfully slow in coming. It was the pace of the times, and many saw no reason why the future should be different from the past. But Newcomen had lived, and Watt, and so Fanny Kemble and George Stephenson knew better.

CHAPTER 9

Boats and Boatmen

DURING the pre-railway age, little public notice was taken of the boatmen who had made industrial expansion possible. They were rough, sometimes dishonest, itinerant, but by no means poor. They may have troubled local landowners and village constables, but they seldom oppressed the social conscience.

On the navigable rivers, craft were sailed when this was possible, masts often being hinged to pass under bridges, and otherwise towed from the banks, in early days by gangs of men called bow-hauliers, later by horses. There were many types: Humber and Yorkshire keels, Weaver and Mersey & Irwell flats, Severn trows and frigates, Norfolk wherries, Thames and Trent barges. Early types often had open holds, with an awning astern supported on hoops, for the crew; later, decked boats with a cabin below, usually in the bow, were introduced.

On the canals there were many shapes and sizes of boat. The best-known type worked on smaller waterways, the narrow boat, was developed by Brindley and John Gilbert from the long narrow craft nicknamed 'starvationers', some 48 ft long and 4½ ft wide, which could work into the Duke's mines at Worsley. Narrow boats were built to fit the 74-ft × 7-ft locks which Brindley had chosen for the central section of the Trent & Mersey as being cheap to build and economical of water in relation to their carrying capacity of 20–30 tons according to depth. The same locks were used on the other Midland canals for which Brindley was engineer, the Coventry, Oxford, Staffs & Worcs and Birmingham, and spread as far north as the Huddersfield Canal over the Pennines from that town to

Manchester, and as far south as Oxford and Worcester. Such boats were usually about 72 ft long and 6 ft 10 ins broad, though dimensions varied a little with local locks. Some canals, notably in South Wales, used slightly shorter and wider boats. Those on the Monmouthshire canals in 1796 were 62 ft 6 ins × 8 ft 10 ins: four of them cost £28 each in that year.

In the days before railways there were few family boats, though the Swansea Canal was exceptional in ruling that 'no Females be allowed to navigate Barges'. Most had cabins, though not the snug floating homes which became familiar on narrow boats later, but some were day boats without cabins, worked over short distances by men who did not sleep or live on board.

Smaller craft, about 20 ft × 6 ft, called tub-boats, were used on the canals which had inclined planes or lifts, most of which could not cope with boats carrying more than five or six tons of cargo. These were plain iron or wooden rectangular tanks, worked chained together in gangs, several being towed by one or two horses.

On the broad canals were various types of barge or wide boat. The short, wide locks of the Leeds & Liverpool, running over the Pennines, took boats 62 ft × 14 ft 3 ins; just south, another waterway over the mountains, the Rochdale, took barges 74 ft × 14 ft 2 ins, able to navigate the Duke's Canal; but at its eastern terminus, Sowerby Bridge, it met the Calder & Hebble with its 57 ft 6 in × 14 ft 2 in 'westcountry' barges, where cargoes from the bigger craft had to be transhipped. In the south of England, the principal broad canals were the Grand Junction, which took barges 72 ft × 14 ft 3 ins, though most of this size only worked at the London end, leaving the rest to narrow boats which could travel through to Birmingham or Leicester without transhipment, and the Kennet & Avon, with 73 ft × 13 ft 10 in craft. Such sizes were maxima – many craft were in fact smaller. On average, a barge might carry 50 to 80 tons according to its size and the depth to which the canal allowed it to be loaded.

Iron was probably first used for a barge when John Wilkinson the ironmaster launched *The Trial* on to the River Severn at Coalbrookdale on July 9th, 1787, though he is said to have built one earlier than this. But iron craft were more expensive than those of wood, and were little used, except for ice-boats. In 1833 the Mersey & Irwell company had an experimental iron sailing flat built for them by Laird's of Birkenhead, the cost, £375, being no more than for a wooden boat. It was so successful that they went over entirely to iron for new flats thereafter, and also for barges to be used for through working to the Rochdale Canal. Iron then began to be used along with wood for bigger craft, and extensively for compartment boats when they were introduced in the 1860s on the Aire & Calder. There were also a few iron narrow boats – the Oxford Canal company ordered two in 1865. Later steel was used.

Company bye-laws usually ruled that a narrow boat should have a man on board to steer – able-bodied and over 18, the Coventry company said – and a boy, whose age should be 12 (certified by the clerk of his parish) on the Ellesmere, 14 on the Stourbridge, and 16 on the Swansea, to lead the towing horse. On the busy lock-flights of the Birmingham Canal Navigations with their perpetual queues of boats, delays through inexperience or incompetence in beast or man could not be allowed: so the company hopefully enacted that no boat should enter a lock without 'an able Horse and able and experienced Hands'.

On the 60-ton barges of the canals round Nottingham, the crew had to be three, one to steer, one to open and shut lock gates, and one to handle the horse. Of these, said the Grantham company, one must be of full age and one at least 16. The colliery owners protested at the manning costs, and although the Grantham and its twin the Nottingham stuck to their point, they later agreed to a steerer of 18, and two others at least 10. Presumably the Grantham acquiesced without altering its own bye-law.

On rivers, when boats were sailed as well as towed, they frequently took the ground and had to be shafted off. Therefore, bigger crews were enforced – on the Trent in 1802, three men and a boy if carrying 18 tons or more of merchandise or coal; presumably less for gravel barges. It was not always wise to be the boy, even on the Mersey & Irwell, where they were company apprenticed (minimum age, 10) and therefore came under the eye of the management:

> James Lister, the Captain of the *Byrom*, be suspended from his employment for a Month in consequence of his savage and brutal Conduct to his Apprentice William Molineux.

Because this company were themselves carriers, we get a glimpse of what employment was like on the flats, partly sailed, partly towed, on a busy and competitive river navigation at the turn of the eighteenth century. In 1797 there had been a threatened strike, a letter having been written to the company from the *Seven Stars*, Warrington, and signed 'The Flatmen', requiring an advance in wages, and threatening to lay up vessels. The manager then went along and reached agreement: men should be paid weekly, 16*s* for the captain, 13*s* the first, and 6*s* to 9*s* the second hand, with bonuses for fast trips, yearly bounties for good behaviour, and cost of living allowances for captains with families in Liverpool or Manchester. Each side was to give a week's notice, and the captain could choose his own hands, subject to the manager's approval. It sounds sensible and rather modern, and worked well for many years.

Canal carrying appears very peaceful and safe. But accidents happened. Men fell into locks and were drowned or crushed between boat and wall, or met injury or death in many unexpected ways. Here is one from the *Gloucester Journal* of February 6th, 1847:

> On Tuesday last, a man, named John Savory, aged about 28, was drowned in the Coomb Hill . . . Canal . . . He was riding on a horse which was dragging a boat laden with hay, along the canal, when the horse slipped and fell with the

deceased into the water, and the ill-fated young man was drowned, before his uncle, who was master of the boat, could render him any assistance.

Most canal craft were built with ordinary open holds, fitted with side and deck cloths for wet weather, but decked boats were provided for valuable cargoes, and a few for special purposes, such as the double-decker sheep boats that worked into Paddington bringing fresh meat to the metropolis.

On many river navigations, especially in the early part of the canal age, barges that could not be sailed were hauled by gangs of men called bow-hauliers. These had to do without a path, plunging along the river's edge as best they could. On almost all canals, and later on most rivers, towing was done by animals walking along a specially built path. Where, for reasons of land ownership or canal engineering, this changed sides, a towpath bridge was provided, often so built that the horse could cross the canal without it being necessary to cast off the towline. On older canals, and those built on the cheap, there was often no towpath under bridges. Some, as on stretches of the Stratford-upon-Avon, Stourbridge and Staffs & Worcs canals, were built of iron in two sections, so that the line could pass through a slot between them. With others, the rope had to be cast off and then picked up again. A few tunnels had towpaths: boats had to be legged, shafted or pulled through most of them, while the towing horses were led over the top along paths which can still be traced.

On the Continent, horse towing was not always from a path. In 1815 Robert Southey noted that at Namur

> we saw a horse in the middle of the river, towing a vessel against the stream . . . a second beast was on board to relieve its comrade, for it is severe work.

Usually the towing horses belonged to the boatmen or the carrying concern which employed them; but on some navigations used by tidal or seagoing craft, such as the Gloucester &

Berkeley, Weaver, and Mersey & Irwell, professional hauliers or trackers provided a service, either under contract to the canal company or by hiring themselves to the boat captain, horses often being changed at regular stages along the route. At Weston Point, where the Weaver Navigation joined the estuary of the Mersey, the horses waited in stables and the men in a clubroom for boats from the Mersey to go up river. The work was hard, though most animals were probably well enough looked after, if only for their value. But some were not. In 1836 the Mersey & Irwell committee, told that horses were 'often unmercifully flogged and otherwise cruelly treated', appointed an inspector to walk the towpath and watch for cases that could be punished 'with the utmost severity of the law'. His work improved matters.

Boats were usually towed by a single horse, sometimes by one or two mules or donkeys. The line ran from the horse's harness to a towing mast rising about 4 ft above the boat's side, high enough to keep the line off the towpath, but not to foul a bridge. In Britain, towing horses spent their nights either in canalside stables, many of which can still be seen, or tethered on the bank. But in the United States, where much longer journeys were worked, many towing horses or mules had their own stables in the forepart of the barge.

Those canals along which coal was a principal traffic, tried to organize a regular flow of boats all moving at the same speed, not allowed to pass or race each other ('all cases of furious driving on the Canal Banks will be punished by a Fine of Ten shillings', the Manchester, Bolton & Bury company enacted in 1844), to be fastened together or to travel rudder first, and given a limited time to unload – 4 hours, the Coventry company optimistically enacted in 1769. When this happened, loaded boats from the collieries to the principal market, in this case Coventry, were given the towpath side of the canal. If such a boat met one going the opposite way, the towing horse of the latter took the inside of the towpath and stopped, so that the towing line fell flat and did not hinder the straining animal

NAVIGATION

FROM THE

Trent to the Merſey.

Notice is hereby given,

That all Perſons not lawfully authorized,

Who ſhall Travel along the

TOWING PATHS

of this Navigation,

Or on the BANKS of the FEEDERS or RESERVOIRS

Will be Proſecuted.

And whereas injuries are frequently done to the *Locks, Bridges, Gates, Wiers, Stops* and other *Works* of this *Navigation* by Perſons unknown; a Reward of ONE GUINEA will be given, over and above all reaſonable Charges, to any Perſon or Perſons who ſhall give Information of any ſuch Offender or Offenders, to be paid upon Conviction.

THOMAS SPARROW,
Clerk to the Company.

JULY 10th, 1800.

SMITH, PRINTER, NEWCASTLE.

18 Towing paths might only be used by those with business on the canal

of the laden boat. At such moments an idle horse might choose to nip his colleague of the other boat, or the boy leading him; so in 1787 the Loughborough company enacted that horses must be muzzled unless they were grazing.

When boats moored for the night, it must not be in a lock or on the towpath side. Sometimes also bye-laws were made against mooring on an embankment, for fear of damage to the bank, or stopping at certain places, for fear of damage to preserved game. For instance, the Cromford company in 1804 ruled that a boat might not stop at night in Crich Chase, or any wood or coppice. If a boat were moored at a public wharf, it was often a rule that no one should sleep on board – otherwise he might have been tempted by the goods stored there. Neither was a cabin fire usually allowed while a boat was near a wharf, for fear of carelessness.

Arguments would develop at a lock between the crews of craft going opposite ways, as to which should use it first. Some companies installed distance posts at equal distances from the lock approaches, the boat to pass the post first having the right to the lock. Optimistic rules were made about how locks were to be worked. Horses were to be unhitched, and boats man-hauled or shafted into the locks to prevent them coming in too fast, stern ropes or straps should be used to check them before they could hit the far gates, and, in case they did, a roller or fender should be fitted to the bow, or a piece of wood held there as a protection. Paddles were not to be dropped without using a windlass, a boat was not to push the top gates open with her bow, nor were boatmen to use the paddles to flush the boat out of an empty lock. But, of course, boatmen in a hurry did all these things, though steerer Thomas Fox in 1802 was prosecuted for going too far, and

> improperly drawing the water at Hatton Locks and, when remonstrated with (by lockkeeper Isaac Cashmore), stripping to fight.

PUBLIC APOLOGY.

LANCASTER CANAL.

I, the undersigned, THOMAS HENRY BURROW, ate of Ellel, now of Gleestone, near Ulverstone, Farm abourer, acknowledge and confess that on Monday Evening, the 8th of March, 1886, I wantonly and foolishly pened a Clough in a Lockgate in Glasson Lockage, which esulted in a waste of water, and for such offence have endered myself liable to a penalty of Five Pounds, or ot less than Forty Shillings, and in default of the ayment of the Penalty, to imprisonment in the House f Correction for not exceeding Six Months nor less han Three Months; I now therefore

PUBLICLY APOLOGIZE

or such offence, and promise that I will not offend gain; I request and hope that the Canal Proprietors vill not prosecute me; and I authorize the printing nd publication hereof at my cost and expense.

Dated this Sixteenth day of March, 1886.

T. H. BURROW.

E. & J. L. Milner, Guardian Printing Works, Lancaster.

19 To waste canal water was a serious offence

Should there be a mishap, and a boat sink, the owner was responsible for the cost of draining the section of canal, and of raising it.

When water was scarce, companies tried to make one lockful serve two boats. They could rule, as the Erewash did in 1796, that narrow boats should pass their broad locks together if within 600 yds of each other, or enforce working turns, that is, a boat going one way had to wait for one going the other before using a lock.

Lock-keepers and toll-keepers were instructed to help boatmen to keep moving – especially if the canal suffered from competition with another route – but also to report drunken boatmen (especially if they were carrying a wine or spirit cargo which they might have broached), and not to trade with them or accept tips, or to rake in the canal for coal they might helpfully have knocked into the water as they went by. These things were easy enough to write in minute books and print in bye-laws, but occasionally one gets a glimpse of the humans behind the words:

> That the complaint against George Cowlishaw be dismissed on condition of his making an apology for his incivility to the Collector's Wife at Beeston Meadow Lock.

Brindley had invented an ice-breaking boat, drawn by mules, in 1766. Later ice-boats were short iron craft with rounded bilges and an iron bar down the centre. With half a dozen or more horses pulling, and a gang of men holding the bar and rocking hard, they worked down the canal to the sound of grinding and cracking ice. But if the ice grew too thick, or if it accumulated in locks to prevent the gates opening fully, or to wedge boats against the walls, then the canal closed, boats were frozen in, pubs' trade boomed, and urgent goods were sent by waggon.

Tunnel working was sometimes a special problem. Where the tunnels were broad, but the boats narrow, as Blisworth or

OXFORD CANAL.

To Lock-Keepers and Boatmen.

TAKE NOTICE, that in consequence of the shortness of water, no Lock shall be drawn off when there is a Boat within sight, or a short distance of it, on its way downwards, but such Boat shall have the use of the Lock before another Boat shall pass upwards.

BY ORDER.

Canal Office, Oxford.

VINCENT, PRINTER, OXFORD.

20 The Oxford Canal company orders its lock-keepers to work turns – as far as possible to make one lockful of water pass two boats

Braunston on the Grand Junction, no regulation of traffic was necessary. Boats, which had to have a lighted lamp or candle, were legged through, men lying on boards set across the deck and pushing against the tunnel sides with their feet. To speed up the passage, professional leggers charging a fixed fee were licensed, and could be hired by the boatmen. Sometimes it was made compulsory to use them, as in the Islington tunnel of the Regent's Canal before tunnel tugs were put on, when a bye-law ruled that barges through the tunnel must have a crew of three men over 18.

In narrow tunnels, one-way working was enforced. For instance, the Dudley Canal company in 1830 allowed boats to enter the west end of Lappal tunnel at 4 AM, 10 AM, 4 PM, and 10 PM; at the east end at 1 AM, 7 AM, 1 PM, and 7 PM, except Sundays. The same company's Dudley tunnel could be entered at the south end at 2 AM, 8 AM, 2 PM, and 8 PM; and the north end at 5 AM, 11 AM, 5 PM, and 11 PM, except Sundays. Lappal was 3,795 yds long, and Dudley 3,172. To speed movement through Lappal, steerers were paid 1*s* 6*d* to 3*s* a loaded boat, so that they could hire legging help. In Dudley a practice grew up of boatmen paying a couple of their number to take several linked boats through while they went home or visited the pub. Naturally the craft moved very slowly, and the canal company did its best to enforce its rules of two persons to each boat. In long tunnels, cabin fires had to be put out.

Later in the canal age, ways of speeding up traffic through tunnels were tried, such as steam-driven endless cables to which boats could be attached. Tugs were found most efficient, either self-propelled or the type which dragged itself along a chain laid on the tunnel bottom. Such tugs could pull a long string of boats faster than they could be legged. They worked on regular timings, and boats were compelled to use them. The first tunnel tug seems to have been that which started work through the Islington tunnel in 1826.

Throughout most of the canal age, canals were closed at night in the interests of lock and toll-keepers, and of safety.

Boats could, of course, move if they wished, but not past toll-houses or the main lock-flights. A chain was put across the canal, or the lock gates padlocked. A frequent practice was to open the canal from one hour before sunrise to one hour after sunset; alternatively it might be done on some such hours as the Erewash company laid down in 1807: 4 AM to 8 PM April

SUBMISSION.

We, the undersigned, Richard Anderton, of 33, Old Lancaster Lane, Weaver, aged 15, Thomas Thompson, of 19, Delacy Street, Tenter, aged 12, Walter Cook, of 92, Old Lancaster Lane, Warehouse boy, aged 12, having been caught bathing in the Preston and Lancaster Canal, and being threatened with Prosecution, do agree to make this Public Submission, pay all expenses, and Promise not to offend again, in consequence of Legal Proceedings being stayed.

(Signed) **RICHARD ANDERTON.**
(Witness) Inspector Mc.Clarnan. **THOMAS THOMPSON.**
WALTER COOK.

Borough Police St
September

21 Bathing was prohibited mainly because the bathers could so easily swim to boats and steal from them

to September, and 6 AM to 6 PM October to March. A few companies recognized the existence of the moon, like the Monmouthshire, who varied the one hour after sunset by saying that boats could run until 8 PM for seven days before and two days after the full moon. Because the Severn traffic was regulated by the times of flood tide in the tidal stretches below Worcester, the Staffs & Worcs, which connected with it, ruled that boats for the spring tides could pass all night, 48 hours before and after the full moon. But some boatmen were always in a hurry, in their own interests or those of the cargo owners. It was therefore recognized that lock-keepers were entitled to a tip for passing boats out of hours, and some companies said what it was to be; on the Erewash 1*s* at Trent lock, at Beeston lock on the Trent, 6*d*.

From the 1820s onwards, however, we see competition increasing, one canal line with another, canal carriers against road transport. This is met to some extent by relaxing the general rules: the Staffs & Worcs in 1816 introduced a £5 annual licence to pass any time, and in 1820 substituted a charge of 2*s* a trip; but much more by having special rules for fly-boats. A fly-boat carried merchandise and light goods; it was supplied with relays of horses, and a crew of four, two to work the craft and two to sleep. Such a boat was allowed to pass others, was given precedence at locks, and could run all night, as well as most or all of Sunday. In theory unhindered, in practice other boatmen were not always willing to give way: in 1825 on the Worcester & Birmingham, fly-boat captains complained that coal barges would not allow them to pass.

Where railway competition was added, however, many canal companies went over to all night working, though without being able to afford shifts of lock-keepers. The Erewash in 1847 made a nominal charge of 1*s* a year for passing boats out of hours, the Loughborough in 1846 ruled that craft should pass all night:

> it being considered advisable to encourage a quick transit of Merchandize by Canals.

In the earlier part of the canal period, boatmen had taken Sunday off. Religious custom suggested it, and laws compelling church attendance were still in existence behind the custom. Some canal companies enacted Sunday closure: the Birmingham in 1770 ruled that no boat might navigate without permission, the Cromford laid down fines for Sunday movement in 1796, the Nottingham prohibited it in 1798. Others, like the Droitwich or the Loughborough, limited closure to the hours between 9 AM and 5 PM. Others again, like the Oxford or the Grand Union, limited their rule to closing public wharves on Sunday, Christmas Day and Good Friday; and some made an occasional exception, such as the Staffs & Worcs in 1811 when they ruled that craft for Bristol spring fair could travel on Sundays.

But, as trade increased, time became more pressingly money, and boatmen began to work on Sundays. In 1795 the Nottinghamshire magistrates received complaints from parishes that barges were being hauled on Sundays, especially at service time, and instructed constables to lay information against the offenders for breaking the law. They also got the Trent Navigation company to close its locks at 8 AM except in emergency. But when local people got a fine inflicted on a boatman near Todmorden on the Rochdale Canal in 1823, the company decided to prosecute those who had brought the action, presumably on the grounds that navigating hours were their responsibility.

In the first three decades of the nineteenth century, therefore, Sunday working became commoner: canals were becoming increasingly congested as trade pressed upon them, and religious feeling was weak. Companies seldom enacted Sunday opening, as the Stourbridge did in 1830 when they opened Stourton locks all Sunday, but they raised no objection to it. Then, about 1839 and 1840, in the sudden uprush of evangelical and tractarian revival, came a change of mind all over the country. In February 1839 the Vicar of Stourport protested against Sunday working, and was supported by the townspeople, causing the canal company to say it was undesirable to stop it, but that the remedy lay with the carriers. In May, 140 boatmen at Shardlow on the Trent & Mersey petitioned that company to close the canal on Sundays: in June the flat captains and hauliers on the Weaver asked for Sundays off; in April 1840 the Lancaster and the Worcester & Birmingham companies circularized others to seek agreement, and in May the Trent & Mersey petitioned Parliament

> for the general prevention of Traffic on all Canals and Railways on Sundays.

But by now railways were being built, and people wanted to travel by them on Sundays. This demand overbore that to prevent general Sunday working.[1] What happened, however, was that many canal companies, who had already stopped their locks on Sundays, continued to do so right to our own days, while their railways competitors worked seven days a week.

With efforts to stop Sunday working went others to bring religion to the boat people, such as those of Mr Burgess, a committee man of the Mersey & Irwell, who in 1819 gave, without comment, a bible and testament to the crew of each flat in the company's service. In 1829, the Paddington Society for Promoting Christian Knowledge among Canal Boatmen and Others, started a little monthly, *The Canal Boatmen's Magazine*, which was to convey

> interesting information relating to the diffusion of religious knowledge among boatmen, dustmen, and labourers at the wharves, particularly at Paddington,

the great canal basin at the end of the Grand Junction Canal. They kept it going for four years.

It looks rather pathetic today, as we read of its Ladies' Association raising money to pay for the Boatmen's Chapel, its Sunday School, its record of tracts distributed, its exhortations against drink and swearing and fighting, its pride in recording the number of boatmen who came to services –

> fifteen canal-men and seventy-five other persons present

[1] For a controversy on a railway, see John Thomas, *The Springburn Story*, 1964.

THE

CANAL BOATMEN'S

FOR THE YEAR 1832.

NEW SERIES—VOL. III.

"In meekness instructing those that oppose themselves; if God peradventure will give them repentance to the acknowledging of the truth."—2 *Tim.* ii. 25.

London:

PUBLISHED BY THE PADDINGTON SOCIETY FOR PROMOTING CHRISTIAN KNOWLEDGE AMONG CANAL BOATMEN AND OTHERS;

And Sold by J. Nisbet, 21, Berners Street; F. Baisler, 124, Oxford Street; or by Mr. Pitts, *Secretary*, 43, Paddington Street.

22 A boatmen's mission at work

—its prayer-meetings on Wednesdays, its shop at the chapel on Monday evenings, when clothing might be bought at cost price, its statement that

> it is very desirable that in conducting the worship of God, due regard should be paid to the devotional part, and the singing led in such a manner as to enable the humblest worshipper to join. Any friend who possesses a knowledge of a few plain tunes, and disinterested enough to use that knowledge gratuitously in this place, would render a very essential service.

Pathetic, maybe, but those who ran the society had more courage and care than many of us would own to possessing.

In 1839 the Oxford Canal company resolved

> that twenty pounds per annum be contributed towards the support of the Minister, Schoolmaster & Mistress of the floating chapel for boatmen,

and in 1841 the Staffs & Worcs company agreed to subscribe £25 per annum to the Boatmen's Pastoral Instruction Society. But when the Rev R. T. Jeffery of the Inland Navigation Society asked the Regent's company for permission to put a floating chapel in the City Road Basin, that secular-minded concern replied that all wharves were let, and all space in the basin was for the tenants' use. Many of these organizations then faded away, not to be revived until George Smith's campaign of the seventies for the Canal Boats Acts, and its accompanying religious literature like *Rob Rat*, aimed at arousing interest in the plight of canal-boat people suffering from low wages and bad conditions partly created by railway competition.

The Weaver Navigation, publicly owned by the county of Cheshire, went further. After an enabling Act in 1840, they quickly built three churches, at Weston Point, where their river joined the Mersey, Northwich and Winsford, and paid for parsons, singers, cleaners and hymn-books. At each place they also established schools for the children of those who worked on the navigation, and paid the masters. The managing committee were closely interested in what they had done; when school attendance fell off at Winsford, they firmly told the schoolmaster

> that the exhibition of a little more kindness to the Children might have a beneficial effect.

This concern for the boat people was no less genuine because it included also a care for society. For it is undeniable that boat people could do with improvement, in the matter, for instance, of theft. Pilferage was a perpetual trouble to carriers and the owners of goods, and a by no means negligible deterrent to carrying goods by water. Most boats had open holds, from which merchandise or drink could be stolen. It was not so much that a sack of sugar or a barrel of brandy would disappear, as that there would be less sugar in the sack at the end of the voyage, and some water in the brandy barrel. Carriers provided some boats with no means of access to the hold except by a hatchway, which was padlocked with a key kept by the captain. The lock itself might be sewn up in a canvas bag to make it less accessible. But the human factor remained.

The canal age ended without any notable use of steam-driven boats to counter locomotives. This was not because of lack of enterprise, for there were many experiments, but for practical reasons. On the canals, mechanical propulsion could only achieve a very small increase in speed over the horse, because the depth and width of existing canals did not permit it, whereas the engine, boiler and fuel took up a considerable part of what was formerly cargo space. Again, although in theory a steam boat could tow several others, this practice only began to show economy on long lock-free stretches. These were few and far between; steam towing proved uneconomical as compared to horse traction even on the main line of the Shropshire Union, which had longer level pounds than most.

From George II to William IV and Victoria; from the Duke of Newcastle as prime minister to Lord Grey and Sir Robert Peel; from the Battle of Minden and the storming of Quebec to the Reform Act and the repeal of the Corn Laws; from the death of Henry Fielding to the birth of Thomas Hardy, Britain's economy had depended, increasingly as the years went by, upon a developing canal system. What was the reality behind such words?

CHAPTER 10

Aspects of the Age

COAL made the industrial revolution, and the need for coal built the canals. The Wednesbury collieries were served by the Birmingham Canal, those at Cannock by the Wyrley & Essington, in south Derbyshire by the Cromford and the Nottingham; in Warwickshire by the Coventry. Moira coal moved on the Ashby; Silkstone on the Barnsley; Clifton and Kearsley on the Manchester, Bolton & Bury; Radstock on the Somersetshire Coal; South Wales production on the Monmouthshire, Glamorganshire, Neath and Swansea canals.

What did this mean in transport terms? In 1845 the Birmingham Canal Navigations carried just over four million tons of goods, of which over half was coal. About 68,000 narrow boats a year swam deep-laden with it behind their sweating, plodding horses into Birmingham itself, or outwards from the mines towards Bromsgrove and Worcester, Stourport and the Severn, or to Warwick and the South. And the same number of boats came back, some lucky enough to get a return cargo in a hold from which the clinging coal-dust had been laboriously swilled, most going empty to queue once again for a load beneath the tips.

Much of this was domestic coal; but over a quarter went to feed local industry in the mining districts, which had 700 steam-engines working near the waterways; a great deal more to iron-making, at a time when there were 1,500 collieries and ironworks on the banks of the Birmingham canals. Transport of coal for iron-making was cheap then: 3*d* a ton for coal, 6*d* for ironstone, any distance over two miles on the whole Birmingham canal system.

It was the same elsewhere. In Scotland, great quantities of coal came down the Monkland Canal with its twin flights of locks, and later the inclined plane at Blackhill, to Glasgow; and when it was hoped that Tyrone coal would prove a major source of Irish supply, the Coalisland and Tyrone canals were built to carry it.

Very small ironworks using charcoal or locally-produced coal, and local iron-ore deposits, could manage with pack-horses, road waggons and horse tramroads, but ironmasters could not create big units until they had a means of bringing raw materials in quantity, and taking away the finished product. The great ironworks of Birmingham and the Black Country, Yorkshire and South Wales, owe their origin to the waterways and the horse tramroads built as feeders to them. The Birmingham Canal Navigations carried great quantities of coal to such works; ironstone too, 451,000 tons of it in 1845, and fluxing limestone, and carried away the finished iron. When supplies of home-produced iron ores became insufficient, then some traffic flows altered; the Glamorganshire Canal began to carry upwards from Cardiff to Cyfarthfa and Penydarren, as, farther along the coast, the Tennant was transporting Chilean copper ores from Swansea in boats that had once been filled with those of Devon and Cornwall.

Cotton was king in Lancashire then. Before 1790 London was the principal port through which cotton was imported, to be carried by canal and road to the North. Thereafter Liverpool went steadily ahead, as supplies from the United States increased. Cotton came from Liverpool docks up the Mersey in sailing flats to Runcorn, and thence by the Bridgewater Canal or the Mersey & Irwell Navigation to Manchester. Some also by road waggon, but comparatively little. Much went beyond Manchester, for instance from the Duke of Bridgewater's Castlefield basin under Deansgate and Piccadilly to the great warehouses of the Rochdale Canal company, and then on to Rochdale itself; or by the Ashton and Huddersfield canals to Ashton-under-Lyne, Stalybridge, Oldham and Stockport.

Much cotton, also, went direct from Liverpool docks by the Leeds & Liverpool Canal, up the great flight of locks at Wigan, and on to Blackburn, Burnley and Nelson. When the Liverpool & Manchester Railway was built, cotton freights were fought for. The railway took some, but much continued to move by waterway until, in 1894, transhipment became unnecessary; cargoes that had been loaded in the southern United States could now come up the Ship Canal to be unloaded in Manchester docks.

Cotton inwards was matched by manufactured textiles outwards, for home consumption and export, from Lancashire, for instance, to Bristol past Middlewich, Market Drayton, Wolverhampton and Kidderminster to Stourport and then down the Severn, to London by the Potteries, the Coventry and Grand Junction canals; and by the trans-Pennine canals to Selby and Hull.

Raw materials for the machines had to be matched by raw materials for the men and women who worked them. A great deal of corn, for instance, came from Lincolnshire, up the Trent and into Yorkshire, and then over the Pennines to the cotton towns; 20,000 tons of it on the Rochdale Canal in 1812, which had become 41,000 tons seven years later. This was a trade that later changed direction, for by the mid-sixties little corn came over the hills; most was imported at Liverpool from the New World. Groceries of all kinds, sugar, currants, wine and a thousand things, also came from the ports, London, Hull, Liverpool, by water to the cotton towns.

Men needed houses to live in, and factories to work in. So bricks came from the brickfields; slates down the Ellesmere & Chester Canal from Wales to supplement what came by sailing coaster from North Wales ports for transhipment at Liverpool or Runcorn; timber over the Pennines from Hull, imported from the northern countries of Europe, 9,000 tons of it over the Rochdale Canal alone in 1818.

Stone, too, for road-making, to meet the incessant demand for better, faster, straighter roads and cleaner, firmer pavements, to carry the mails, the passengers and the goods of an England whose production was rising every year. Much of Manchester's stone came from the great Dove Holes limestone quarries near Buxton, which were worked by the Peak Forest Canal company – probably some quarter of a million tons a year. From the quarries it was carried by the company's horse tramroad down to Bugsworth, which later generations call Buxworth, to be shipped on the canal past Marple locks and aqueduct, Hyde and Dukinfield to the city.

When the stone was limestone, our ancestors got through enormous quantities, for iron-making and road construction, but also for land improvement, especially of waste. The early nineteenth century was a time when men saw it as necessary to bring every acre into cultivation that could be made to yield a crop; it was the other side to the argument about the corn laws. And nothing was as good as lime. Limekilns were built near the quarries if the coal slack needed to burn it was available locally; if not, then limestone would often be taken towards the coal, as that from the quarries at Llanymynech where the Ellesmere Canal met the Montgomeryshire was carried to be burnt at Vron near Ruabon. Quantities were considerable: in 1814, 32,500 tons of lime and limestone came down the Cromford Canal from the Crich and other quarries; in the same year 26,000 down the Rochdale. Trade had improved since 1812, when the Huddersfield Canal committee had complained of

> a great depression in the demand of Lime, for the improvement of Waste Lands on this line of Canal.

The Peak Forest company worked hard to market their lime and limestone, in ways that have a modern touch. They offered to pay part of the cost of new kilns, and cheap canal tolls for their first years of operation; grants and loans also for building the boats to carry it. They made agreements with other canal companies which quoted a single rate for the cost of the stone, toll and freight, and got the product sold for some thirty miles around, at Runcorn and Leigh, Bolton, Bury, Marsden and beyond Rochdale. They arranged cheap bulk delivery contracts with turnpike trustees, and sought to subsidize

> a Treatise (which may be afforded to be sold for One Shilling) on the nature, effects and proper Application of Lime as a Manure.

It all added up to a great inland waterway trade, on waterways that in England and Wales had grown from a mileage of some 1,400 in 1760 to 2,224 in 1790, 3,691 in 1820, and 4,023 in 1850 (see the table in Appendix I). Contemporary estimates put the total at about 30 million tons in England and Wales in 1830, and it grew for some years after that. We have figures for some of the canals serving South Lancashire and running towards the Pennines or the South: in 1838, 2,220,468 tons on the Leeds & Liverpool Canal; 514,241 on the Ashton-under-Lyne and 442,253 on its connexion the Peak Forest; some 950,000 on the Bridgewater and Mersey & Irwell together, and about a million on the Trent & Mersey which joined the Duke's Canal at Preston Brook. Also serving the area were the third Pennine Canal, the Huddersfield; the Manchester, Bolton & Bury, the Sankey Brook Navigation, the Weaver and the Ellesmere & Chester to Ellesmere Port. Between them they probably accounted for another two million. Even after deducting overlapping figures for goods carried on more than one canal, and remembering that part of the tonnages do not refer to the Lancashire area, these are not negligible figures for days when railways were few, and roads only bulk carriers of heavy goods for short distances.

The arguments for the Rochdale Canal's bill of 1792 summarize what our ancestors hoped for from building canals: that the new waterway would give Manchester, Rochdale, Oldham, Burnley, Colne and Halifax more efficient transport of hardwares from Sheffield, Birmingham and Wolverhampton, pottery from Birmingham and Etruria, glass from Stourbridge and Gainsborough, hops, cider and fruit from the cider counties, timber, wool for manufacturing, coal for power and household use, stone, limestone, the corn and produce of many counties, and imports from overseas brought in at Liverpool or Hull.

Even the country canals, that wound their peaceful way through the fields, carried substantial tonnages: in the same year of 1838, 31,130 tons on the canal from the River Wey to Basingstoke; 47,295 on that from the sea to Chelmsford; 40,421 on the Melton Mowbray Navigation to that town, with some of it going on by the Oakham Canal to Rutland; 61,899 on the Wilts & Berks, winding from Abingdon past Wantage, Swindon, Calne, Chippenham and Melksham towards Bath down the Vale of White Horse and the valley of the Bristol Avon.

What did such a country canal carry? The Montgomeryshire Canal (Eastern Branch) ran from Garthmyl (where it joined another from Newtown) to join the Ellesmere Canal by the limestone quarries at Llanymynech. It ran through an area that was being improved agriculturally, and therefore limestone for lime-burning (and road-making), and slack coal for the kilns, were major traffics westwards, together with coal and general merchandise and groceries. Eastwards the boats returned mainly with building materials for eastern houses: timber, building stone, and slates.

Here are figures for the trade westwards to Garthmyl as the canal played its part in rural development.

	Limestone	*Slack coal*	*Coal*	*Merchandise*
1806	24,082	6,757	5,534	611
1814	44,592	11,560	11,863	1,075
1824	51,241	15,882	23,377	2,489
1834	53,634	16,578	20,159	5,385
1844	45,155	15,674	19,852	5,872

Eastwards from Garthmyl

	Timber	*Building Stone*	*Slates*
1806	1,605	578	1,217
1814	2,883	759	2,682
1824	4,839	1,746	2,121
1834	1,222	2,886	1,361
1844	2,872	2,373	1,202

Once the pioneers, John Ashton on the Sankey Brook, the Duke of Bridgewater between Worsley and Manchester, Josiah Wedgwood on the Trent & Mersey, had pointed the way, established industrialists quickly saw the advantage to them of the new, more efficient and cheaper means of moving heavy loads. They therefore took a prominent part in promoting canals that would pass near their premises, often becoming subscribers and committeemen. Such main lines constructed, they would then build private canal branches, or perhaps tramroads, to connect them to their property. In Shropshire in 1788, Richard and William Reynolds of Ketley ironworks were founder committeemen of the Shropshire Canal, to which Richard subscribed £6,000 and William another £1,000; they also built the private Ketley Canal, nearly two miles long, from their works to join it. Again, Matthew Fletcher, working the Wet Earth and Botany Bay collieries at Clifton on the bank of the Irwell, in 1790–1 built the 1½-mile long Fletcher Canal to give his coal access to the proposed line of the Manchester, Bolton & Bury Canal, of which company he was an original committeeman, and to which he had subscribed £1,500.

Once canals were open, men built new factories close alongside the waterways, with wharves or canal-connected basins where their boats could load or unload. When the Sankey Brook Navigation was cut to it in the early 1760s, St Helens was a village. The canal had been planned to develop the local collieries, but some ten years later, a plate-glass works was built, and then one for copper smelting; thereafter the neighbourhood quickly developed. One has only to glance at old plans of such towns as Birmingham or Stoke-on-Trent to see industry crowding the banks of the Birmingham canals or the Trent & Mersey, or to read contemporary newspapers to note how any sale advertisement for an industrial property told how far it was from a canal. As for collieries, every quarter-mile away from the water diminished their value.

The siting of industrial enterprises alongside canals went on happening well into the railway age: for instance, in the building of chemical works along the banks of the River Weaver and its associated Weston Canal to the Mersey, or at Widnes, to have access to coal brought down the Sankey Brook Navigation from the St Helens field. Later, businessmen were to choose rail-connected sites, and later again to look for good road access, free of traffic congestion. But today on Britain's most modern waterways, and extensively along new Continental canals at Ghent and elsewhere, one sees factory owners seeking the canal's banks in order to participate in the cheap transport it affords.

When canals first replaced road waggons, a spectacular drop usually took place in the price of such bulk commodities as coal, the transport cost of which was a large part of the price: in Birmingham from 15*s*–18*s* to 4*s*–7*s* a ton. But another factor soon appeared. In road transport days, a town or district drew its coal from the nearest colliery – choice was too expensive. The first canals widened choice by providing supplies from a colliery district – from the whole Wednesbury area as soon as the first Birmingham Canal had been opened in 1769. But, as canals joined up into a system, so coal from each district moved towards that from another, until some sort of balance was achieved between the various local prices, made up of pit-head prices plus canal tolls and freights, and the differing kinds and qualities of coal.

This balance was, of course, always unstable as canal companies subsidized the penetration of other companies' districts by offering especially cheap tolls for coal going beyond their territory, often being supported by coal-owners who offered reduced pit-head prices for such cargoes. This equalization of the market is well seen in the area between Oxford, Reading, Bradford-on-Avon and Cirencester, which was served by the Oxford Canal bringing Warwickshire and Moira coal southwards, the Kennet & Avon and the Wilts & Berks that from the south Gloucestershire and north Somerset fields, and the Thames & Severn carrying coal brought down the Severn from Staffordshire, or up the tidal Severn from the Forest of Dean or South Wales. So the consumer was given a range of coals to choose from at varying prices, industrial productivity and personal pleasure increased and transport costs tended to fall.

The same thing happened with corn. When this could move about the country by water, or be brought easily from the ports, the danger of local famines receded. In spite of the strains in the market caused by the French and Napoleonic wars, references in canal records to local famines die out soon after 1800. The big inland towns could be fed only after the producing area had been thus extended. As these grew, so food had to be brought from farther away, by waterway and also by coasting vessel and road waggon. Not only corn, but all other foodstuffs. For instance, the canal freight revenue of apples carried from the cider counties to Manchester by way of the Trent & Mersey Canal alone in 1791 was £6,000 per annum.

A rough and ready way of showing the fall in transport costs that was going on is to look at some representative cost per ton figures taken from canal records. In detail these are unreliable, because they take no account of variations in the length of haul or changes in the type of cargo, but in total they show the tendency well enough (see table overleaf).

Nearly every town of any importance before 1841 was either on the sea or an estuary, with facilities for receiving and sending goods by coastal shipping, or on a navigable waterway, or both. Of seventy principal towns in Great Britain,[1] only one,

[1] *Abstract of British Historical Statistics*, 1962, Population and Vital Statistics, Table 8, p 24.

Luton, with 3,000 people in 1801 and 6,000 in 1841, was on neither. In Appendix II, I have listed the towns that were on inland waterways only, with their populations in 1801 and 1841, and the date when they were given canal communication.

		Representative costs per ton		
	Year	*Tons carried*	*Tolls*	*Tolls per ton*
			£	*s d*
Rochdale Canal	1812	199,623	17,153	1 8·6
	1822	408,967	30,942	1 5·2
	1832	556,711	43,236	1 6·6
Cromford Canal	1803	155,776	5,780	8·9
	1817	227,211	8,210	8·0
	1833	286,539	9,675	8·1
Regent's Canal	1824	454,256	25,684	1 1·6
	1834	624,827	28,930	11·1
Montgomeryshire	1806	42,850	2,745	1 3·4
(*Eastern Branch*)	1816	52,799	3,494	1 3·9
Canal	1825	99,548	6,095	1 2·7
	1834	102,007	5,134	1 0·1

It is not easy to visualize how such growth could have taken place in the inland towns without waterways. Horse tramroad trucks, and road waggons, each limited to about two tons, could hardly have done it alone. One may speculate fantastically whether, the need being so great, the locomotive would have been developed sooner; but then the conditions of railway building would have been more expensive, for the extent to which railway construction material was carried by canal is often ignored. What if it had all had to be moved by road?

It is impossible to give anything like an accurate estimate of the number of people who worked in the inland waterway business in its prime. George Smith, the reformer, in the 1870s, reckoned the canal boat population at 100,000. This was almost certainly an over-estimate for his day, although it is likely to be roughly correct for, say, 1841. If we take 80,000, however, and add 20,000 to include the canal companies' and carriers' staffs, we have 100,000 as a reasonable guess at the number directly concerned with waterways. At this date boatmen were still earning good money, and kept wives and children in houses ashore; the family boat was a product of later rate cutting against the railways, especially on the narrow canals of the South and Midlands which had less ability to fight back than the bigger and more economic waterways of the North. So employment on the boats was mainly of men, and on the wharves and in the lock-houses, a woman in charge was a rarity. In 1841 the male population of England and Wales, over 14 and under 70, was 4,691,600. Perhaps we may guess, then, that 1 male in every 47 was engaged in the inland waterway business. Very many more, in factories, mills, collieries and works, in quarries and fields, depended to some extent for their livelihood upon the canals' puddled clay, busy locks, sweating horses, and crowded wharves.

The conduct of canal companies was regulated by the provisions of their own Acts. These had often been passed only after hard bargaining with local property owners, industrialists and merchants had produced a compromise that then strictly governed the exact route to be followed, water supplies to be taken, and maximum tolls that could be charged. If changes later became necessary, these had to be authorized by amending legislation, which often proved difficult to get without giving interested people the chance of reopening settlements made at the time the original Act had been passed. Many a company abandoned a bill to allow them to raise tolls because their construction costs had exceeded estimates, or to enlarge their powers, or to build a branch canal, rather than allow older privileges to be questioned.

The granting of an Act was then thought of as involving certain public responsibilities, and conferring certain rights. Because land and road improvement was considered as a benefit to the locality as well as the individual, the Act usually provided that manure and roadstone for use in parishes penetrated by the canal should be carried free of toll when water supplies were satisfactory, a local subsidy to agriculturalists and road authorities that must have becn missed when the railway age came. On the other hand, the canal being regarded as a kind of property, it was thought right that Parliament, in authorizing a second canal which, by competing with the first, might alter the value built up under the protection of its original Act, should provide redress. Hence the compensation tolls often granted to an older canal company when a newer one wanted to connect with it: charges they could levy on traffic passing the point of junction. These tolls had an inhibiting effect on expansion of the waterway network: some projects ceased to be economic propositions after a compensation toll had been allowed for. When railways came, a few early agreements between canal and railway companies did provide for payments as compensation for competition, but these soon died away in face of a change in Parliamentary opinion away from these older ideas.

Whereas, later on, railways were governed not only by their own companies' Acts, but by a growing body of general railway legislation, this was not the case with waterway companies in the canal age. Apart from the Registration (Barges) Act of 1795, a wartime measure to list all waterway craft, their crews, and their customary journeys that almost at once proved a dead letter, an abortive bill of 1797 to levy a tax on goods carried by inland waterways that Pitt dropped in face of the storm of opposition it aroused, and the Constables Act, 1840, which empowered Justices to appoint special men, usually employees, as constables to guard company property on the application of a canal company, there was no general legislation before 1845, when two Acts were passed with the intention of helping canals more effectively to compete with railways. These had been encouraged and prepared by a group of canal representatives who had originally met at Osborne's Hotel, Adelphi, London, on May 6th, 1844. They had sounded out Mr Gladstone, President of the Board of Trade, who was favourable to legislation, and had then prepared a bill which was later split into two. One of these Acts specifically authorized canal companies which adopted it to become carriers – though some already were. The other enabled them to vary tolls; that is, to quote different tolls for passing the same distance on different portions of their line, instead of a strict mileage rate, and to make working arrangements with, or to lease themselves to, other canal companies.

Living conditions on canal boats were not regulated until the Canal Boats Acts of 1877 and 1884, or tolls brought into general order until a series of Acts in 1894, which in turn derived from power given to the Board of Trade under the Railway & Canal Traffic Act of 1888.

High rates of canal dividend are often quoted as examples of the excessive fruits of a monopoly situation. But on the whole this is not so. Our canal building ancestors lived in an age of inflation, as we do, but differently arranged. Those who held shares in a waterway built before the French wars of the 1790s had begun, were in a similar position to those who subscribed for Leyland Motors or Woolworth's shares when they were first issued.

In Appendix III will be found some representative figures in four groups. The first comprises canals completed by 1790; the second a smaller list of some started and finished in the 1790s; the third a selection of canals also begun in the 1790s but which, for various reasons, were completed after 1799 and before 1816; and lastly, some begun after 1800 and finished at dates up to 1835. The average costs of construction per mile are:

	£
Group I	3,323
Group II	4,256
Group III	9,725
Group IV	12,972

Too much must not be read into such figures, but they do show the trend. Allowances must be made for broad canals costing rather more per mile than narrow, and for rural canals costing less than those through mining and industrial towns. Really urban canals, built partly through already built-up areas, were, of course, much more expensive still; for instance, the £42,875 per mile of the Birmingham & Warwick Junction, opened in 1844 through part of Birmingham, or the £74,737 per mile of the Regent's Canal through London, completed in 1820.

Shareholders in canals in Group I who had retained or inherited original shares found themselves part-owners of companies with comparatively small amounts of issued capital: £70,000 for the Birmingham Canal, £135,000 for the Staffordshire & Worcestershire, £177,148 for the Oxford. Those who courageously bought the Mersey & Irwell from its original shareholders in 1779 for £10,000 were equally fortunate. These figures can be compared to those for two major Group III companies, the £983,080 capital of the Kennet & Avon, or the £1,160,000 of the Grand Junction. The same kind of comparison can be applied to borrowings, usually incurred early in a company's life in order to finish construction; on cheap canals interest charges were small, on expensive ones a good deal higher.

As traffic grew, the old companies benefited disproportionately, both because their original cost had been low, and because newer canals which had joined them introduced fresh business. Hence high rates of dividend upon small issued capital. Nowadays, of course, most of these companies would have issued bonus shares and reduced dividend rates, but in those days the Birmingham company was exceptional in doing so.

I have taken the year 1830 to list in Appendix IV the dividends, converted to percentages, of the same companies as were quoted in Appendix III. The average figures are as follows: some comments will be found in Appendix IV.

		per cent
Group	I	48·1
	II	13·12[1]
	III	5·19
	IV	·33 (for the three canals finished by 1830)

By 1830 Britain's economic growth probably required that canals should be extensively rebuilt and modernized to suit somewhat changed traffic patterns, and that they should be complemented by a railway and improved road network to provide an efficient and interlinked system. But what situations require, and what actually happens without benefit of hindsight, are very different. As we shall see.

[1] See note in Appendix III.

CHAPTER 11

To Modern Times in Britain

THE horse plodded along the towpath of the Glamorganshire Canal. Abercynon locks had been passed, and the waterway ran high on the western shoulder of the Taff valley, Merthyr Tydfil a few miles ahead. Below, and beyond the river, a horizontal line along the hill marked the tracks of the Penydarren tramroad, a horse-operated plateway serving local ironworks. Evan Williams at the tiller was keen sighted; he could see a feather of smoke appear on the line, and a four-wheeled, tall-chimneyed iron contraption move ponderously down, behind it a string of small trucks, some filled with iron, others with men. Faintly the rattle and thump of its passing was carried to him across the valley, before the barge-horse's clopping hoof-beats again became the only noise he heard. The date was Tuesday, February 21st, 1804; the engine Richard Trevithick's, first locomotive to propel itself along a railway track, a phenomenon and a portent.

Yet change came slowly. Not until 1841 was the Taff Vale Railway opened up the valley, nor till 1898 was the upper section of the canal closed, defeated at last by the locomotive. Today, the mineral lines that honeycomb the Welsh valleys are themselves contracting before the omnipresent motor lorry; contracting, but not yet defeated. For heavy traffic, roads yielded to canals; they to railways, who have now lost predominance once more to roads, and to pipelines also. In transport there is no finality – only transition, new and old and older overlapping, gaining, losing, as needs, direction of flow, techniques, dictate.

Trevithick's engine was indeed a portent. The cloud no

bigger than a man's hand was first clearly seen in the autumn of 1822, when the Mersey & Irwell company saw a notice of intention to go to Parliament

> for a Rail or Tram Road from Manchester to Liverpool.

The Liverpool & Manchester Railway opened in 1830, and caused the competing Bridgewater and Mersey & Irwell Navigations to cut their rates. In so far as any generalization is true, it is probably correct to say that early railway companies saw themselves as concerned mainly with carrying passengers and light goods, and regarded bulk traffic as likely to clutter up their lines with slow-moving and relatively unprofitable goods trains. But not for long. Their need for more revenue over which to spread overheads, manufacturers' need for quicker transport, combined to drive them towards merchandise and minerals carrying, the basis of inland water transport. In turn, carriers put on more fly-boats to speed up the movement of merchandise, but canals had little passenger business. Therefore, while railways could spread their overheads over passenger and goods receipts, most waterway companies were entirely dependent on the latter.

Because of the speed with which industrial Britain was developing, people needed to get about the country cheaply and efficiently, while the speeding up of merchandise traffic meant that firms need hold smaller stocks; therefore the same capital would finance a greater quantity. Waterways had taken from 1760 to 1840 to add 2,600 to the 1,400 that already existed; railways took about seventeen years from 1830 to equal this mileage. Then they shot ahead. In the one year 1846, railways were authorized, and subsequently built, equal in length to the whole waterway system.

Naturally, therefore, new works tended to be built near the growing railway system, sidings being provided as a matter of course, just as older works were canalside, served by private waterway branches and docks or covered warehouses. Similarly, older collieries, mines and quarries used water transport; with some exceptions near Birmingham, newer ones had rail connexions. For some time, therefore, the way a firm received its raw materials depended upon the situation of its supplies; the way it dispatched them, upon its own and that of the recipient.

Then slowly the older collieries acquired rail sidings, or perhaps were worked out and closed, and older works became connected to the newer system, or, if industrial congestion made this impossible, moved to new sites. New businesses went to rail; water could only retain some of the old.

From the beginning, almost all railways organized carrying departments, offering a single rate and a single responsibility. Collection and delivery services were soon added. Independent carriers now became agents of the railway, or offered special services. Though very few canal companies were legally prohibited from carrying goods on their own or others' water, in practice they regarded themselves as making a track available for others to use, on the analogy of turnpike trusts as providers of roads. Thus, in the critical thirties and early forties, when railway managements were busily seeking business, offering services backed by a company of known standing, canals were dependent upon the energy, or lack of it, of carrying firms, some large but many small, few with the seeming solidity of railways, and many carrying also by road and rail, and therefore not fully committed to water transport. Carrying by canal companies was not formally authorized until 1845, or borrowing to establish carrying departments until 1847.

A few canal concerns were already carrying; some others now set up carrying departments, that worked not only on their own, but on neighbouring waterways, like the Grand Junction, whose craft ran beyond the limits of their own canal at Braunston near Rugby, to Birmingham, Leicester and the Trent, or the Aire & Calder, or the Trent Navigation. Many never did, relying on independents or the carrying departments of their neighbours. As with railway trucks, many firms owned canal boats and did their own carrying, and some railway companies, having bought control of canals, themselves started to carry, as they did on the rails. Only one, however, the Shropshire Union, did its best to create a carrying monopoly.

At nationalization, some carrying departments were absorbed, and carriers taken over; most remained independent, so that today, though British Waterways has carrying fleets of its own, most carrying on its waters is done by independents. On the Continent the older ideas have prevailed: waterway authorities maintain the track, but others use it.

Canal companies were badly placed to compete with railways on long-distance routes. The short profitable mineral canals of the north Midlands, like the Nottingham or the Erewash, continued to carry much of the traffic for which they had been built, though at lower tolls; the Birmingham Canal Navigations went on linking Black Country works, collieries and railway basins. But between Manchester and City Road basin in London, for instance, there were seven canal companies on one route, eight on another, no less than eleven on a third; on the competing railway line, by 1846 only one. This situation gave railways the initiative: given a sight of a promising traffic, they would quote a rate. But before canal companies could quote competitively, they had to consult, and their committees, often meeting at widely separated towns a few times a year, to ratify. By that time they had often lost it. If, on the other hand, small canal companies did not act in concert, but each changed its tolls in its own interest, the result was a medley of changing rates that alike confused carriers and those wanting to ship goods. Some waterway amalgamations took place: the Shropshire Union, the Birmingham Canal Navigations, what later became the Sheffield & South Yorkshire emerged, all under railway control, and an extended but independent Grand Junction. But far too many small companies struggled on to nationalization, in many cases keeping alive less by waterway tolls than by revenue from property and sometimes water sales. Therefore it was long-haul canal traffic that suffered most. Some was retained at very low tolls, or by making rates agreements with competing railways; most was lost, the canals being driven back upon short-haul, local traffic where they had special advantages.

Other handicaps faced canal companies who, regarding their position as a safe one, had distributed too much money in dividends and ploughed too little back. Few canals had the equivalent of a railway double track: most locks were single, and many narrow boat tunnels were the same. Some doubled them, as the Trent & Mersey company did at Harecastle tunnel, and down their flight of locks to the West, but many did not. The Staffordshire & Worcestershire's key line southwards past Wolverhampton to the Severn remained through the canal age much as it had been built in the 1770s, while shareholders divided a regular forty per cent. Single locks meant also stopping traffic for repair. Canal companies got used to grouping their annual stoppages for a week, but each one was an invitation to traders to move to rail.

After allowance has been made for the exceptions, the picture at the end of the canal age is one of waterways too small, too slow, too fragmented, indeed too old-fashioned to stand up to the Victorian rush towards railways, a rush whose impetus still comes to us from the weekly files of the *Railway News*, *the Railway Gazette*, *Herapath's Railway Journal*, and many others of the day. If canals were to continue alongside railways – and, of course, coastal shipping and roads – as a means of internal transport with its own national value, there was no hope, in Britain of the 1840s, of modernization, rebuilding, re-equipment, by private enterprise, preoccupied as that was with railways. Only the State could have done it, and the British Government, after putting a cautious toe into the whirlpool of railway control and even nationalization, quickly took it out again. State action to maintain an independent waterway system was not then even considered.

It is in the light of individual canal companies competing with individual railway companies that the purchase of many waterway concerns must be seen. There were some moves towards local alliances of waterway companies, on the line of canal from the Derbyshire coalfield towards London, for instance, and round Birmingham, but shareholders seldom remained proof against a tempting offer. Certainly those of the Birmingham Canal Navigations did not. There were no comparable alliances of railways against waterways: each was dealt with separately.

Railways bought control of canals for a number of reasons: to get rid of competition and enable rates to be raised to a profitable level; to avoid damaging Parliamentary opposition to their bills; to use their land for their lines; occasionally to penetrate a neighbouring company's territory, as did the North Staffordshire Railway with the Trent & Mersey. Occasionally, canal companies took the initiative: as early as 1831, the Manchester, Bolton & Bury decided to convert their canal to a railway, though the railway line opened in 1838 did not in fact use the waterway, which remained open. The little Liskeard & Looe Union in Cornwall built a parallel railway alongside the canal, operated it, and allowed the canal to become disused; and the Lancaster Canal company for some years leased the competitive Lancaster & Preston Junction Railway.

Railway-controlled canals now formed parts of many, though not all, through routes. Because it was usually railway policy to prefer carrying by rail, unless enemy territory was to be invaded, obstacles were put in the way of cheap canal through tolls, in spite of all Parliament could do, and so long hauls fell away still more. The condition of such canals tended to deteriorate, and this discouraged traffic. One can put on the credit side of the ledger that the railways had the stability to keep their canals open, and therefore they survived to nationalization to be used for pleasure cruising. Given the finances of the Macclesfield Canal, for instance, it would almost certainly have closed in late Victorian days if it had remained independent, like the Melton Mowbray Navigation from near Leicester to that pleasant town.

By the 1870s and 1880s a confused picture was becoming clearer. A few waterways were doing well: the Aire & Calder, energetically managed, secure in the coal trade of its Yorkshire mines, the Weaver in Cheshire under its trustees, adding chemicals to its staple traffics of salt and coal; the Sharpness New Docks company, owning the Gloucester & Berkeley Ship Canal, and having recently acquired a narrow-locked waterway from Worcester to Birmingham; the River Lee in London, and the Grand Junction route to Braunston towards Birmingham; the Birmingham Canal Navigations, busy with local business. Railways were working others hard, for reasons of their own. But many long lines were faltering, notably the three over the Pennines, the Shropshire Union from Wolverhampton to the Mersey and its branches into Wales, and the line from the Grand Junction to Leicester and the Trent. That river's own navigation was in difficulties; so were the Commissioners of the Severn and the Thames. Minor canals were closing: part of the Grand Western, in 1867; the Chard in 1868; by 1869 the Lower Sussex Ouse and the Itchen were disused; in 1871 the Wey & Arun and the Torrington; in 1875 the Parrett Navigation and the Baybridge; in 1876 the Coombe Hill and the Bedfordshire Ivel; in 1877 the Melton Mowbray; in 1878 the Sleaford; in 1881 the Herefordshire & Gloucestershire; in 1888 the Arun.

In the eighties, encouraged by the Parliamentary battle for the Manchester Ship Canal Act, and then by the building of that great waterway, there was a healthy revival of interest, especially in big modern waterways, which culminated in 1909 with the great *Report* of the Royal Commission on Canals and Inland Navigations. Until this period, one can read British canal records endlessly, without finding a single reference to those abroad. A few British canal engineers had gone overseas to teach, like Thomas Telford to the Göta Canal in Sweden; but none went to learn. It was only now that Britain began to realize that whereas inland water transport was at home generally regarded as obsolete, abroad waterways were being

CARLISLE CANAL

NAVIGATION.

NOTICE

Is hereby Given,

THAT by virtue of "The Provisions of the Port Carlisle Dock and Railway Act, 1853," the above Navigation will be CLOSED, on and after the First Day of August next, when the water will be run out of the Canal for the purpose of converting it into a Railway. It is expected that the Railway will be Open for Traffic before the end of the present Year, of which, however, due notice will be given.—Dated the 12th Day of July, 1853.

By order of the Committee of the Carlisle Canal Company.

WILLIAM WARD,

JULY 14th, 1853. CLERK.

CARLISLE: PRINTED AT THE OFFICE OF A. THURNAM.

23 The railway takes over

rebuilt, modernized, extended to provide a heavy transport network side by side with railways and roads.

From the eighties onwards, references to Continental waterways became more and more frequent, until they formed part of the evidence upon which the Royal Commission made its recommendations. There seemed a real chance of a policy for modern canals; solid talks were given, solid books written, some of which have never been bettered. In 1888 the Society of Arts held a two-day conference on Canals and Inland Navigation. In opening it, Sir Douglas Galton said:

> We in England have practically ignored our canals for the last fifty years; but when we turn to the Continent of Europe and to America we find that the Governments of France, Germany and Belgium have been much more alive to the importance of water communication; these Governments have acquired, and are improving, their principal water routes with very great advantage . . . in the United States . . . on many important lines of communication it affords an important check upon the rates of traffic carried on railways.

In 1890 came J. S. Jeans' 500-page *Waterways and Water Transport in different countries*, advocating State control; in 1896 L. F. Vernon-Harcourt's *Rivers & Canals*, a two-volume book on waterway engineering, still the standard work on the subject in England; J. E. Palmer's *British Canals: problems and possibilities*, published while the Royal Commission was sitting, recommended State assistance, though not nationalization; U. A. Forbes' and W. H. R. Ashford's *Our Waterways* of 1906, perhaps the most constructive single book ever written on the British waterway problem, which, regarding navigation as a branch of water conservancy, advocated nationalization, waterways then to be entrusted to a Central Water Authority. Lastly, in 1906 also, J. A. Saner, the lively engineer of the Weaver Navigation, gave a lecture 'On Waterways in

Great Britain' to the Institution of Civil Engineers. He was clear that only the Aire & Calder and his own waterway had really moved with the times; of the Aire & Calder he said that under its energetic engineer W. H. Bartholomew it was

> equal to many Continental canals of recent construction.

Saner advocated public ownership and management, whether by local authorities, like his own, or by the Government. He was supported by Sir Edward Leader Williams, engineer of the Manchester Ship Canal, who said Continental countries had built canals to carry raw materials from their ports, and manufactured goods back,

> at rates which were very low compared with the charge in England. In France, Belgium, America and Germany – England's competitors – the canals had greatly improved trade. In many cases the work had been carried out by the State ... it was self-evident that Britain ought not to be behind in a matter of that kind.

And along lines that such men as these had suggested, the Royal Commission reported in 1909, recommending nationalization of the main waterway routes, and their enlargement to a mainly 300-ton barge standard on the river sections, and 100 tons on the canal portions, but with some parts running up to 750-ton standard.

The trouble about their report was, perhaps, that it offered the prospect of a State canal system competing with private enterprise railways. If the former paid, it would be at the expense of the latter; if it did not, the necessary subsidy would be equivalent to an artificial lowering of traders' rates. It was, therefore, immensely unpopular with the railway interest, for whom E. A. Pratt spoke in his widely-read books *British Canals* and *Canals and Traders*, and unattractive to Government. Moreover, Irish troubles loomed; Commons struggled with

Lords over the Parliament bill, and in the background impended the dark shadow of war with Germany. Meanwhile there had been craft development. Iron had to some extent replaced wood. Steam tugs, sometimes themselves cargo carriers, pulling one or several craft, were common on the bigger waterways, and steam-driven narrow boats were in use. Soon diesels would replace them. In Yorkshire W. H. Bartholomew had introduced the 40-ton pan to carry coal, at first six being pushed by a tug (the first successful push towing) and made to curve round corners by steering chains running the length of the train and controlled from the tug, later with the tug conventionally towing up to nineteen. On the big rivers like the Severn and Trent, horse towing died out before tugged or self-propelled craft, or occasionally a tractor working along the towpath, though on some broad, and even narrow canals, it survived to our day.

The war put canals under Government control until 1920. By then their position was worse, for old habits of water transport had often been broken during the war when boats ceased to be available for cargoes because their crews had gone to fight, and costs had risen faster than revenue.

After the war the many railway companies were regrouped as four. There was talk also of canal grouping; further studies were made, but little happened except major improvements on the Trent, financed by Nottingham Corporation, and the amalgamation of canals on the London–Birmingham and, later, the London–Derbyshire coalfield route into the Grand Union Company, which obtained Government help under legislation for the relief of unemployment for widening the Braunston to Birmingham length of narrow canal to that of the former Grand Junction to London, most of the work being finished in 1934.

In some ways the most interesting result of the Grand Union amalgamation was the subsequent diversification of the new company's interests. They not only acted as toll-takers on their canals and at Regent's Canal dock; they carried, by water and road (they even got power to carry by air); provided stevedoring services; started a shipping line to the Continent; developed warehousing; built up an estate business; and found running a lido at Ruislip profitable.

But once again war interfered. Much the same happened as in 1914; boatmen were called up (though some women were recruited), and old transport habits were broken. Soon after it had ended, and before the tottering canal companies could recover what was left of their prospects, most of the country's waterways were nationalized as part of the Docks & Inland Waterways Executive, and later the Waterways Sub-Commission, of the British Transport Commission. There were exceptions, notably the biggest of all, the Manchester Ship Canal, controlled by Manchester Corporation, and its ancillary, the Bridgewater; the Thames; the Yorkshire Ouse, partially controlled by the Corporation of York; the Nene and the Yare.

The commission was a transport authority. It spent some £6 million on considerable improvements to the major waterways, notably lock mechanization and modernization of handling equipment, but found the rest of its canals an embarrassment. However, it built a small hire-cruiser fleet, put on passenger-carrying day and holiday boats, and encouraged pleasure cruising, pending final decisions on the future of the smaller waterways. Two inquiries and fifteen years further on, by which time commercial traffic on smaller canals had fallen away still further, the commission was replaced for canal purposes by the British Waterways Board. It was not everything the men of seventy, sixty, fifty years ago had sought; but it was nevertheless on the right lines.

In the Board's thinking, the major waterways, together with the docks, warehousing and estate sides of their operations, had a viable commercial future. The rest had not. After a detailed survey of every waterway, the Board concluded that, even if every other canal it possessed were maintained in its cheapest form, whether for pleasure cruising, as a water channel only, or abandoned, the cost to the nation would be some £600,000 per annum. For another £340,000 per annum all those useful for pleasure cruising could be maintained, and this sum could be expected to fall as the popularity of canal boating grew. This sum, and the detailed figures on which it was based, were presented to the Government in December 1965, in *The Facts about the Waterways*. The Ministry of Transport then set about seeking the views of all concerned bodies and people. These supported the conclusions the Board had reached, and in September 1967 the Minister of Transport announced the Government's acceptance of the principle of a pleasure cruising network, and a subsidy adequate to maintain it.

On the commercial side, however, the situation remains unsatisfactory. Britain has a number of major waterways other than the Ship Canal: the Aire & Calder, leading from the Humber to Leeds and Wakefield, now enlarged to 540-ton barge standard; the Sheffield & South Yorkshire from the Humber to Rotherham, for which the British Waterways Board prepared an enlargement scheme which has been rejected by the Ministry of Transport in favour of rail; the River Trent to Nottingham; the Weaver in Cheshire, able to take fair-sized ships; the River Severn downward from Stourport to Gloucester, and the Gloucester & Sharpness Canal thence to Sharpness; the Exeter Canal; the Thames and the Lee. A case can be made for improving and enlarging most of them, a case also for extending some of them further inland, and for linking them by seagoing barges or barge-carrying ships with the Continental waterway system that begins across the Channel at Antwerp and Rotterdam, and spreads across Europe.

But whereas there are many decision-makers who would admit that certain British waterways, for instance the Aire & Calder, are necessary and useful, they do not go on to conclude that inland water transport should be generally and normally examined along with roads, railways and pipelines as a means of moving goods about the country, especially in bulk containers, and that barge-carrying ships or seagoing barges are alternatives to roll-on, roll-off ships. There are national main road and railway authorities: there is no national waterway authority, charged with preparing a national water transport policy to be integrated with the others.

No one will pretend that in this country waterways are likely to play as great a part as they do in western Europe. But the country that built the Manchester Ship Canal need not automatically assume that it is always more sensible to carry goods by road or rail to a port than to bring the ships inland; or to use a lorry or a railway truck rather than a large capacity seagoing or ship-carried barge, able to penetrate without transhipment to the four corners of Europe.

Of the two main modern uses for canals, goods carrying and pleasure cruising, one is now accepted by Government and people. Half the job is done; the other half is yet to do. If it can be done as well, this country will benefit socially and economically. On one side a new canal era begins: on the other, we still make do with the old.

CHAPTER 12

The Great Ship Canals

When seagoing ships were under a hundred tons or so, they could often work up navigable rivers to ports within the tidal limits, and higher if the rivers were deep and wide enough, or locks could accommodate them. They still can, up such natural rivers as the Rhine, the Amazon or the Plate, or through the locks, for instance, of our own Weaver.

Most ship canals were the artificial equivalent of such river channels. They could be very small, taking craft such as Yorkshire keels which normally worked inland, but made short coasting voyages. Such was the Louth Canal, dating from the 1760s and 1770s, with locks only 15 ft wide, that ran for 11 miles through Lincolnshire to the town of Louth. Rather bigger but much shorter, was the Newry Ship Canal in Co Down, 1 mile long, with an entrance lock 130 ft × 22 ft, finished in 1769 (it was extended and enlarged in the 1840s), or the 1⅜-mile long Ulverston Canal in Furness, opened in 1796, with its 95-ft long and 22-ft wide entrance lock. Bigger still, the 5-mile long Exeter Canal, rebuilt in 1827, the entrance lock 131 ft × 30 ft 3 ins, which enabled craft of nearly 400 tons to reach Exeter from Turf in the estuary of the Exe. Biggest in the canal age, the 16¼-mile long canal, also opened in 1827, which could admit 600-ton vessels through its 163-ft × 38-ft entrance lock, at Sharpness on the Severn, and carry them to Gloucester docks. There were similar examples abroad: for instance the Terneuzen Canal, once again opened in 1827, that brought ships upwards from the River Schelde to Ghent.

In more modern times, much bigger examples of this kind of

canal have multiplied. There is our own Manchester Ship Canal, opened in 1894 at a cost of £14,350,000, 36 miles long from Eastham on the estuary of the Mersey to Manchester with five locks 600 ft × 65 ft to take large liners, and 179 acres of water space at Manchester. A few examples on the Continent are the great Dutch Noordzee Canal from Ijmuiden to Amsterdam, the Trollhätte Canal in Sweden to take 2,000-ton ships out of the Göta river and lift them through six locks to Lake Vänern, or the recently reconstructed Saimaa Canal, to carry 1,600-ton ships for 26 miles and through eight locks from the Gulf of Finland to Lake Saimaa. An interesting such canal is the short Lindö from the Baltic to Norrköping in Sweden, opened in 1962, because it was excavated not by digging, but by blasting both the surface clay and the rock beneath at one operation (the biggest single blast used 4,700 holes loaded with 55 tons of dynamite), and then removing it all with a water-borne bucket-dredger.

In North America, the finest is the St Lawrence Seaway, part of the astonishing Great Lakes route, first made practicable for seagoing ships in 1848. A complete reconstruction was finished in 1959, and opened by the Queen and President Eisenhower. Now ocean-going ships climb the seven locks, five Canadian and two American, of the Seaway that starts at Montreal, rising 225 ft in 182 miles of river and canal to Lake Ontario. Thence they can by-pass Niagara Falls by using the Welland Canal, 28 miles long with eight locks, that was first opened in 1829 and has been three times rebuilt, to reach Lake Erie. They can then navigate the St Clair and Detroit rivers to reach Lakes Huron and Michigan and so pass the Sault Ste Marie locks into Lake Superior, 600 ft above sea level, and work to Duluth, 2,342 miles from the Atlantic.

But the short-cut ship canals are the most romantic – among these our own small Crinan and Caledonian, the Kiel, Corinth, Suez and Panama – canals which slice through land obstacles to bring nearer the seas beyond. They have stormed and conquered obstacles nature has put in the way of man's adventuring on the sea, those tongues of land connecting land masses that seem to have been laid there as a battle-challenge.

The first mentioned was also the smallest; the Crinan Canal through the Mull of Kintyre, 9 miles long with 15 locks from Ardrishaig to Crinan, opened in 1801. It was built by private enterprise, partly for philanthropic reasons as a means to the development of western Scotland, but before long had to be taken over by the State. Today, the coasting 'puffers' still use it, and many yachts also.

The Caledonian was State-built, begun in 1803 to link the western and eastern seas by cuts to join together the lochs of the Great Glen from Fort William to Inverness, opened in 1822, partly reconstructed in the 1840s, and reopened in 1847. Its original purpose had been to speed the passage of sailing ships, and especially sailing warships, between the east and west coasts of Britain during the wars with Napoleon, by enabling them to avoid the long and turbulent voyage round Cape Wrath. But before the canal could prove its worth, steam had begun to replace sail, and today this waterway, most beautiful to pass along as it curves between the Highland hills or traverses Lochs Ness, Oich and Lochy, carries more yachts than fishing or commercial craft.

The Kiel Canal, across the peninsula that separates Denmark from the mainland, is older in history than either. One can indeed look back to the Vikings, who near Schleswig used horses to drag their ships across on rollers, or the medieval route from the North Sea by way of the Eider river to Flemhude, and then by land transport to the Baltic.

The Danes built the first waterway, the 25-mile long Eider Canal, in 1784, to link the upper Eider river at Rendsburg to Kiel and so eliminate transhipment to road waggons. It was big enough then, with six locks 100 ft × 25 ft; Malthus called it 'the famous canal' when he saw the first lock upwards from Kiel fifteen years later, and said the number of craft that used it was increasing every year.

> The Canal is ten feet in depth, & admits vessels of three hundred tons burden, so that there is no necessity, except in very particular cases, for unloading & loading again . . . The tolls are very trifling, only 8 shillings and fourpence for every vessel . . . This money is laid out in paying the persons who attend, & other expenses relating to the canal. The return to the King of Denmark is solely from the Customs, & these are rather high.

The Danish–Prussian war of 1864 transferred ownership of the canal to Prussia, soon to become the major partner in the German Empire, which began to expand her Navy, and needed a ship canal. It was opened in 1895 from sea to sea, 61 miles long, and soon outgrew its Naval origin to become a great trading route which now carries a greater tonnage than Suez and Panama combined.

Smaller in dimensions, less busy, but incomparably spectacular, the lockless 4-mile long Corinth Canal, opened in 1893, runs in a deep cutting with smooth, towering rock sides through the steep ridge that divides the Gulfs of Corinth and Aegina.

And so to Suez, begun and opened as a contribution to world unity, available under the canal company's regime to the ships of all nations in peace and war, but now become a political counter and a means to divide mankind. It is as well that the statue of de Lesseps no longer stands at Port Said, to see how greatly men can fall below man's ideals.

Like the Corinth, but unlike the others, Suez is a sea-to-sea canal without locks. Its construction was indeed held back until engineers had proved untrue the earlier conviction that the Red and Mediterranean seas had different heights. The ex-diplomat, Ferdinand de Lesseps, conceived it, dreamed of it, studied the physical, economic and above all the political conditions of building it. He had luck, for it was luck that had given him friendship with Prince Saïd when he was consul at Cairo in the 1830s, though it must have been honest purpose and deep conviction that decided Saïd, when he came to the Egyptian throne in 1854, to give his old friend the canal concession.

It is a romantic story – of de Lesseps, dreaming, thinking, planning a Suez Canal, but seeing little chance of getting it started. Then, in September 1854:

> I was busy amongst bricklayers and carpenters, superintending the addition of a storey to Agnes Sorel's old manor house, when the postman, bringing the Paris mail, appeared in the courtyard. My letters and papers were handed up to me by the workmen, and my surprise was great on reading of the death of Abbas Pacha and the accession to power of the friend of our youth, the intelligent and warm-hearted Mohammed Said. I hurried down from the scaffolding, and at once wrote to congratulate the new Viceroy. I told him that I had retired from politics, and should avail myself of my leisure to pay my respects to him.

Just two months later, in a tent in the Egyptian desert at sunset,

> I fet strong in my composure and self-control at a moment when I was about to broach a question on which hung my whole future. My studies and reflections on the Canal between the two seas rose clearly before my mind, and the execution seemed to me so practicable that I did not doubt I should be able to make the prince share my conviction. I propounded my scheme without entering into details . . . Mohammed Said listened with interest to my explanations . . . He brought forward several objections with considerable intelligence, to which I replied in a satisfactory manner, for he said at last: 'I am convinced; I accept your plan, We will talk about the means of its execution during the rest of the journey. Consider the matter settled. You may rely on me.'

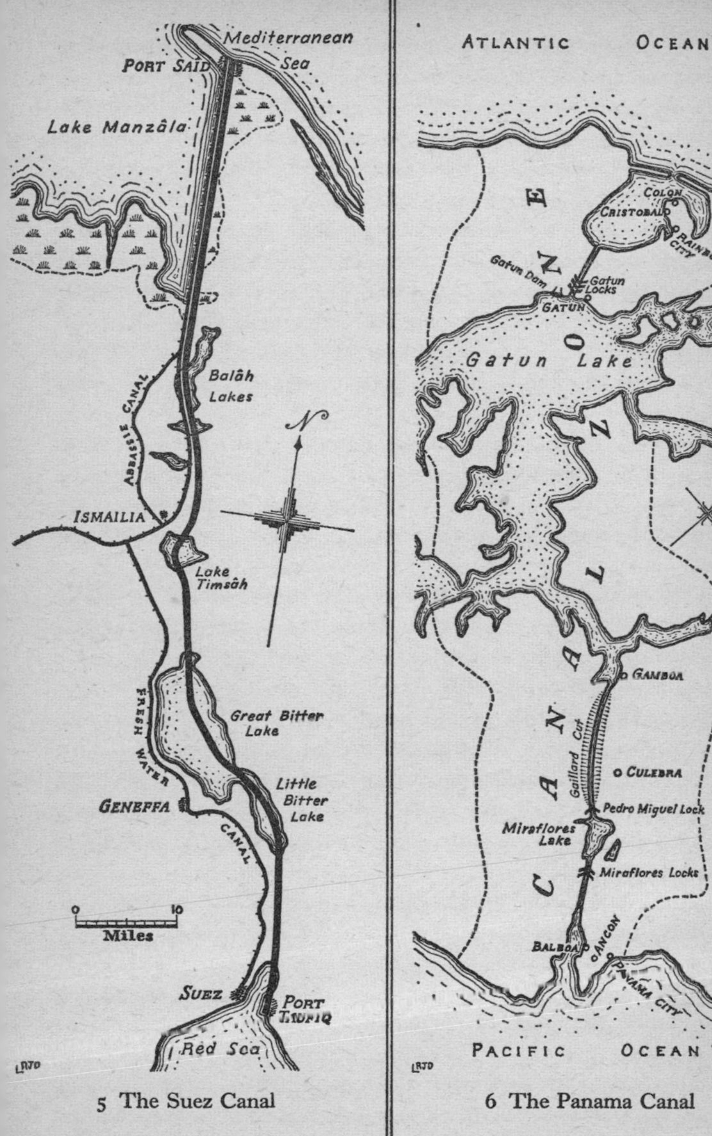

5 The Suez Canal

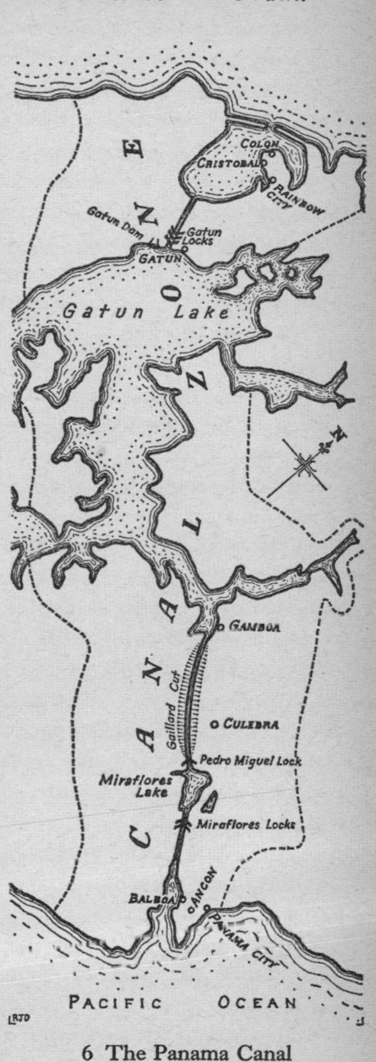

6 The Panama Canal

De Lesseps was no engineer, but he was magnificent both as organizer, public-relations man and politician, one who could recruit, house, feed and pay the great labour force of 20,000 men required to build the 100-mile long canal through the sand; who could raise the money needed from shareholders predominantly in France, but from all over the world as well, and also steer his successful way through great power rivalries as each country assessed the changes in world forces his success would bring about. It was done at last, and on November 17th, 1869, the great canal was formally opened by de Lesseps' cousin and firm supporter, the Empress Eugènie of the French, traversing it in the paddle-wheeled Imperial yacht *Aigle*.

Because the bottom and sides do not need to be puddled to keep them watertight, the canal's depth and width can be increased by dredging, and this has been steadily done ever since its opening, so that the cut de Lesseps made is now much bigger.

And Panama. Through the thin mountainous spine of Central America de Lesseps, an old man now, but famous and triumphant from his success at Suez, planned to cut a 50-mile long lockless sea-level canal. He was at work from 1881 to 1889, but two factors in particular defeated him: the mosquito, bringing malaria and yellow fever to his workmen in the days before the cause was known; and the slippery strata of the central mountain range, which caused more and even more material to slide into the immense cutting he sought to drive from coast to coast.

If he could have brought himself to admit that he had been wrong to try for a sea-level cut, and had substituted locks, he might well have succeeded in building both the world's greatest waterways. But he was too old, too committed, too confident, and so failed in more ignominy than he deserved.

The Americans took over, the mosquito was defeated, and a canal with locks largely on a new line was built and opened in

1914. Now it is reaching its capacity and once again there is talk of building a sea-level canal, as de Lesseps sought to do ninety years ago. But medical knowledge, the ability to move great masses of earth, and maybe the use of underground atomic explosions to do the necessary blasting promise a better result.

CHAPTER 13

The Age on the Continent

THE great rivers of the Continent of Europe, and many lesser ones, have always been navigable after a fashion. Certainly by the 1300s, probably earlier, staunches had been built for barges to pass the millers' weirs in the Low Countries, France, Germany and Italy.

Several ancestors produced the European canal age. In the Netherlands and Italy, existing drainage channels were enlarged and made navigable. In Germany, the first waterway over a summit was constructed between 1391 and 1398. It was called the Stecknitz, and rose 12 ft from Lake Mölln south of Lübeck, then ran level for 7 miles, partly in cuttings, and entered the River Delvanau, itself made navigable by the building of staunches, to end at Lauenberg on the Elbe. The Duke of Milan's engineer, Bertola da Novate, built the first canal to overcome a considerable gradient by means of pound-locks in 1452–8, the 12-mile long Bereguardo Canal, which rose 80 ft. Finally, Leonardo da Vinci invented the mitre-gate for locks, the type still used on smaller waterways, about 1485, after he in turn had become Milan's ducal engineer. This simple design of lock encouraged men to make rivers navigable and build canals, so that by 1600 it was known and adopted throughout Europe, from Britain to Sweden and Brandenburg.

Canal construction spread slowly, in Italy, Germany, Flanders, the Netherlands. It is to France, however, that we look in the early seventeenth century for the first major work. This was the Briare Canal, built to rise from Briare on the Loire up the valley of the little River Trezée, then to Rogny, and so down the Loing valley to Montargis, where there was connexion to the Seine. Hugues Cosnier was chosen engineer after the job had been advertised by 'son de trompe, cry public et affiches'. He surveyed the line, chose the levels, and had his proposal agreed by the Royal council in 1604. Sully provided soldiers to do some of the labouring, and he and King Henry IV visited the canal under construction. But in 1611, with the work three-quarters done, the King was dead, Sully had resigned, and both money and incentive to continue were lacking. It was 1638, and Cosnier himself was dead, before three promoters got letters patent from Louis XIII to take the canal over and finish it. It was opened in 1642, 34 miles long, rising 128 ft from Briare and falling 266 ft to Montargis, with 41 locks having rises up to 14 ft, taking craft some 80 ft long by 15 ft beam. Included in the line were several staircases, probably the first to be built, one falling 65 ft with as many as six chambers, and a double towpath, barges being towed by men on each. The King offered a prize for the first boat through.

The waterway was a great success. Before long the company was receiving 13 per cent on its capital, and carrying 200,000 tons a year. Today, reconstructed, the Canal de Briare is busier still, part of a canal line from the Seine to the Rhône. With Cosnier's biographer, we too can in imagination say: 'Saluons, au passage, la mémoire de ce grand ingénieur!'

The Briare was followed fifty years later by another Loire-Seine canal, this time from Orléans to Montargis, and then by the greatest yet, the Languedoc, to join the Mediterranean to the Atlantic. It began as a vision of the energetic and ambitious King Francis I, who in 1516 discussed possible routes with Leonardo da Vinci, and then had a line surveyed up the Aude to Carcassonne, and to the Garonne near Toulouse, the route that was indeed chosen later. But it was too big, too technically difficult, too expensive a scheme for such times, and, though examined and discussed occasionally, it had to wait until 1662 for the engineer, Pierre-Paul Riquet; the organizer, Colbert; and the grandeur-seeking king, Louis XIV.

Work on the canal was started late in 1666, with Riquet in charge, twelve resident inspectors supervising sections of the work, and some 8,000 navvies, and built with an efficiency many later engineers would have done well to copy. Riquet died in 1680, a few months before the line was opened in 1681 from Sète on the Etang de Thau, giving access to the Mediterranean, to Toulouse, 150 miles long, with numerous locks, including a staircase of eight near Béziers, many aqueducts, and a 180-yd tunnel that is the first I know of built for a navigable canal. One astonishing feature was Riquet's provision of water to the summit level: a feeder 27 miles long to bring water from the River Sor at the Montagne Noire, with another, over half as long, feeding a reservoir in the valley of the Laudot at St Ferréol, its dam 105 ft high and over ½ mile long.

The Languedoc (now called the Canal du Midi) was the first modern canal: in its time astonishing, so that Voltaire, chronicling the wonders of Louis XIV's reign, could write:

> Le monument le plus glorieux par son utilité, par sa grandeur, et par ses difficultés, fut ce canal de Languedoc qui joint les deux mers.

It is still open, still useful, and still displays to the world the engineering skill and organizing brilliance that built it 300 years ago. The Continental canal age had begun; moreover, as we saw in Chapter 2, the Canal du Midi influenced the Duke of Bridgewater, father of British inland navigation, which in turn was the chief inspirer of North America. As railwaymen look to the Liverpool & Manchester Railway as the progenitor, and George Stephenson as the magician of his age, so may we to the Canal du Midi, and to Pierre-Paul Riquet.

The waterway system of Europe expanded slowly throughout the eighteenth century, rapidly in the first half of the nineteenth. Then the pace of new construction slowed down, but still continued far more energetically than in Britain, while effort was also given to the enlargement and improvement of what already existed. Expansion took two main forms; the improvement of rivers by building locks and cutting bypass canals to difficult stretches; and the building of new

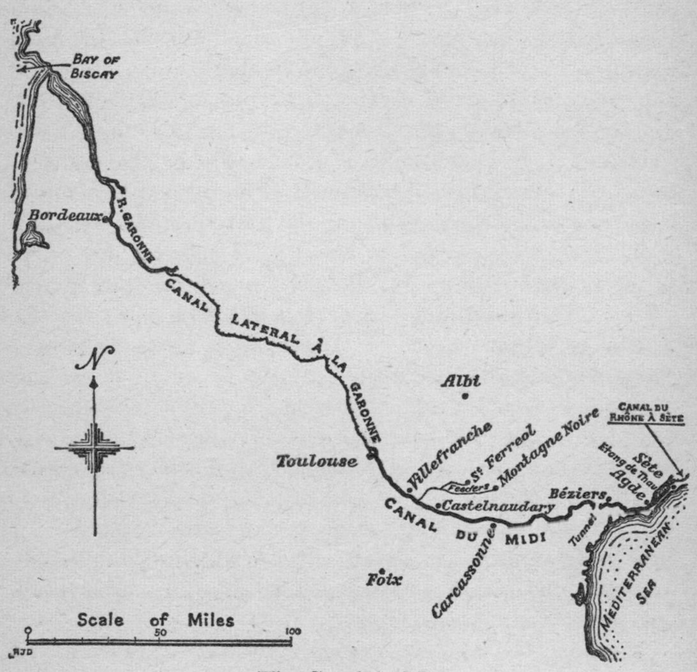

7 The Canal du Midi

canals. Slowly a network of interconnected waterways was created, spreading onwards from the ports to the industrial areas, the capitals, and the interiors of France, Belgium, the Netherlands, and Germany, over which barges passed with little regard for frontiers. A small system based on the Po was created in Italy; the Danube was given some artificial tributaries, and farther east, canals were built in Russia, and rivers made passable.

Here one can only take a few examples of what was done.

King Ludwig of Bavaria, a man who sought greatness, fired, so it is said, by reading about the Erie Canal, decided to build a waterway to link together the River Main, itself a tributary of the Rhine, at Bamberg, and the Danube by way of Dietfurt on the River Altmühl, and so fulfil the dream of enabling barges to pass from the North to the Black Sea. What is more, the company he formed built it in the nine years to 1846, with 100 locks, at a cost of £1½ million. But it could only take 130-ton barges, and these, though often too large for the then unimproved rivers at each end, proved too small to withstand the competition of the new railways. The British shareholders in the canal company, over one-third of the total, saw little return on their money. The Ludwigs Kanal was just usable in the Second World War, though it is now derelict, but the vision remained, and today excavation of a great new waterway, to take 1,350-ton barges, is creeping from Bamberg towards Nuremberg and the Danube.

In Scandinavia a glance at the map shows that the sea passage from the North Sea to the Baltic is dominated by the narrow passage of the Kattegat. In the days when Denmark levied tolls on all ships passing, the advantages of a canal across Sweden were obvious. A look at the map on p. 173 shows the problem clearly. To connect the Göta river above Gothenburg to Lake Vänern; that to Lake Vättern; that to little Lake Roxen; and that to the Baltic at Söderköping.

Thanks to one dedicated individual, Count von Platen, supported by an enthusiastic canal company, the Trollhätte Canal was opened in 1800 from the Göta river to Lake Vänern, by eight steep locks that climbed the rocky cliff past the great cataract. That done, von Platen obtained the King's agreement to continue the canal, and in 1808 wrote to Thomas Telford, at work on our own Caledonian, to ask him to help. Telford agreed, went to Sweden, took to von Platen at once, spent six weeks surveying the whole length to decide the 58 lock sites, and then retired to von Platen's manor to write his report before returning home.

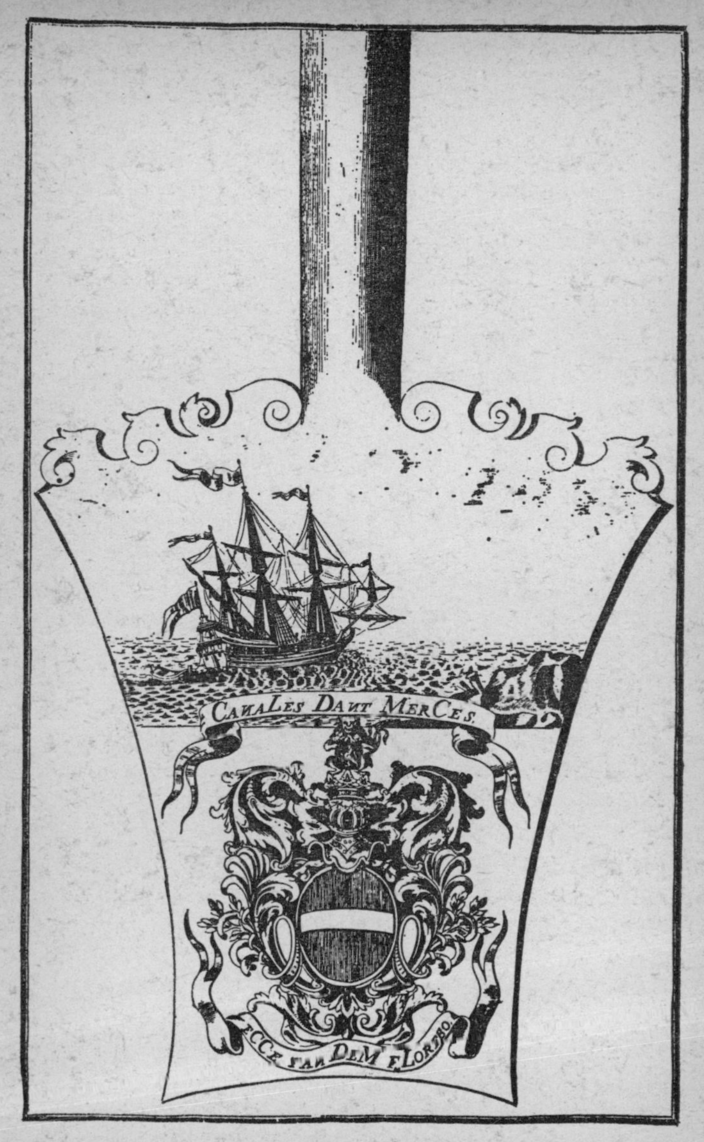

24 One side of an ornamental shovel used when cutting the first sod of the Canal du Louvain in Belgium (then the Austrian Netherlands) on February 9th, 1750

The work took until 1832, 55 miles of canal, 133 miles of lake, with the Göta river and Trollhätte Canal making up 238 miles from sea to sea. Steamers were allowed in 1855, and today, as the visitor sits on the deck of the *Wilhelm Tham*, the only remaining steamer in the Göta Steamship Company's fleet, as

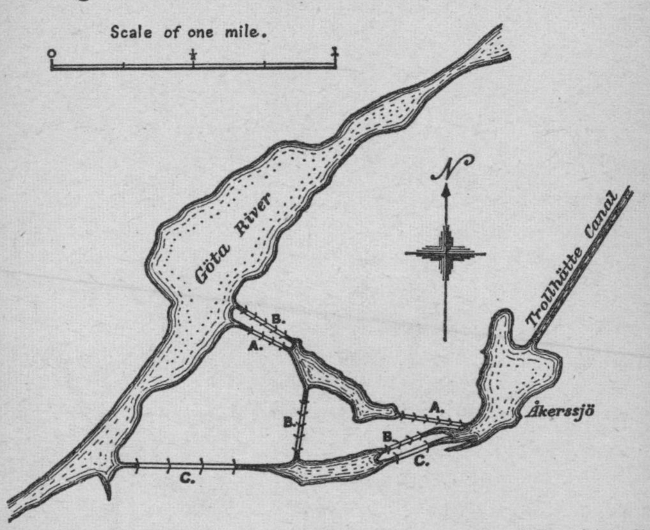

8 The three successive sets of locks built round the falls of Trollhättan, on the Trollhätte Canal in Sweden between the Göta river and Lake Vänern. They were opened respectively in 1800, 1844 and 1916

it moves quietly along what is perhaps the most beautiful canal in the world, he notices at the locks a resemblance to the Caledonian Canal, and wonders for a moment if the Highlands have been transported to Sweden. And in the canal company's headquarters at Motala, he can still see the lock drawings Telford sent over to von Platen, and admire copies of his report in English, transcribed by Swedish clerks in impeccable copperplate, even to the careful copying of the engineer's signature.

In Russia, about 900 miles of canals were built to link

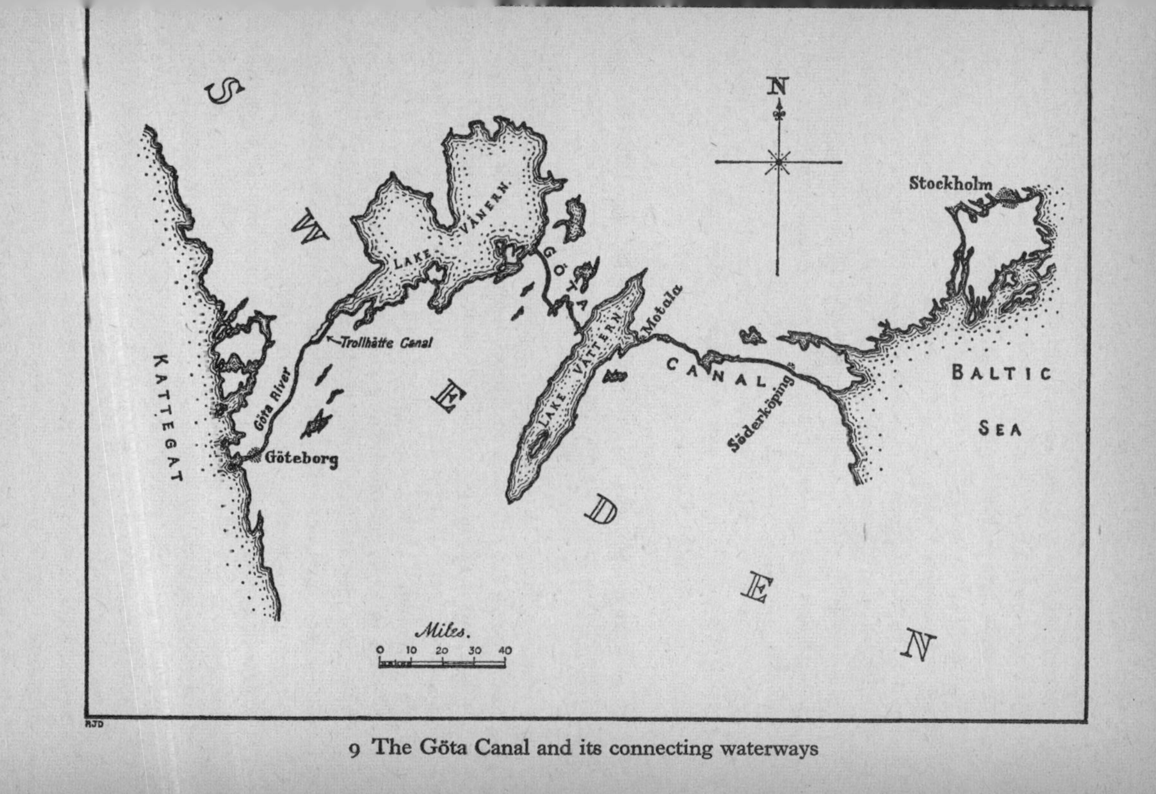

9 The Göta Canal and its connecting waterways

navigable rivers or sometimes lakes, only open in summer, winter transport being by rail or sledge, but carrying great quantities of raw products to the south and west, to bring back manufactured goods. By them, the Baltic was connected to the White Sea, by two routes to the Black Sea, and to the Caspian by way of the Volga. This great river, 2,000 miles long, yet with a fall of only 663 ft, offered with its tributaries some 7,200 miles of navigation. The coming of steam revolutionized long-distance water transport over such distances as these. In the 1800s, for instance, an observer noted that over 600 steamers worked on the Volga. In 1885 the British Vice-Consul at Odessa on the Black Sea recorded an extraordinary voyage by a 150-ton ship, the *Samuel Owen*, which had arrived there from Baku on the Caspian

> via the Volga, the Marie system of canals, Lake Onega, River Neva, and thence round Europe.

In 1966, 400,000 tons of cargo moved on this Baltic–Caspian route (with its branches to Moscow and to the Black Sea) to and from Iranian ports alone, saving 2,700 miles between Germany and Iran compared with the sea passage via the Suez Canal.

And in 1967 an oil drilling rig, the *Chazar*, was towed in three sections by tugs from Holland to Leningrad by way of the Kiel Canal, and then by Russian rivers and canals to the Volga, and so to the Caspian Sea and Baku.

For a typically industrial canal grouping on the Continent we may take north-east France and the contiguous part of Belgium, where the waterways that link the Lys, Scarpe and Schelde carry the products both of the coalfields and the many works that have arisen near them. These waterways created the areas they serve, and even now, as the traveller passes quickly through by car or train, he will see beneath him barges moving, barges moored, barges locking through, still busy, and waterways still developing.

In Chapter 5 we noticed the opening in 1875 of the vertical canal lift at Anderton between the River Weaver and the Trent & Mersey Canal in England, designed for hydraulic operation by Edwin Clark for Edward (later Sir Edward) Leader Williams. The basis of his design was used by the firm of Clark, Standfield & Clark on a short industrial waterway, the Canal du Centre, from Mons to the Canal de Charleroi à Bruxelles, opened in 1895, which uses four such lifts to take 400-ton craft, raising them a height of 211 ft in the course of 4 miles. These in turn inspired others in Germany and Canada, and still do their work.

The Canal de Charleroi à Bruxelles itself carries the products of the busy Charleroi area on the Sambre to Brussels and Antwerp. Originally built for 300-ton craft, with a kilometre-long tunnel and 38 locks, its complete reconstruction to 1,350-ton barge standard was finished in 1968. Now with no tunnel and with 10 locks, its most exciting feature is the inclined plane at Ronquières, where a 5 per cent slope nearly a mile long gives a total rise of 220 ft, perhaps the greatest canal structure in the world.

The struggle to the sea can be exemplified by Amsterdam. Though with 40 ft of water in its roadstead, there was only 10 ft farther out in the Zuider Zee; thus large vessels could not reach the city fully laden. In 1819 the Great North Holland Ship Canal was begun, 50½ miles long, 124½ ft wide at surface, 20¾ ft deep, running up the north Holland peninsula past Alkmaar to what was then called Nieuwediep, and is now Den Helder, with deep water close inshore. The great canal was completed in only six years, and Amsterdam was at last accessible for seagoing ships of about 800 tons.

The greater part of it was later enlarged to 2,000-ton standard, with a shortened connexion to Amsterdam via Zaandam; but that also being far too small for the great city's needs, the Noordzee Ship Canal, straight across the peninsula for 15 miles to Ijmuiden on the North Sea, was built between 1863 and 1865, with Sir John Hawkshaw as consulting engineer.

Its water level is only 42 centimetres above *low* water in the North Sea, so that normally it is below sea level, with, at Ijmuiden, locks which include the largest on any canal, 1,320 ft by 165 ft. Along this canal, straight across the low Dutch countryside, the biggest ships can now reach the city.

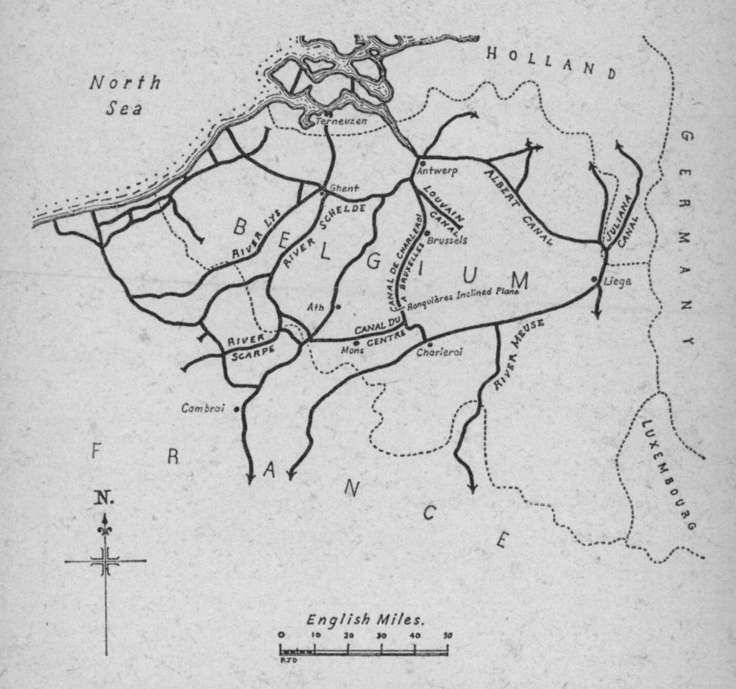

10 The industrial canals of west Belgium and north-east France

For our last example, let us choose one of the three great French long-distance waterways. There is the Canal de Bourgogne, 150 miles long, which was started in 1775 but not opened throughout until fifty years later. It begins on the River Yonne, a Seine tributary, and climbs an astonishing 923 ft to its summit 1,240 ft above the sea, passes a 2-mile tunnel, and falls 612 ft to the Saône for Lyons and the Rhône to

TWO ASPECTS OF CANAL LIFE: (*right*) a decanter, engraved with CCC and a horse towing a canal boat, from the boardroom of the Coventry Canal Company; (*below*) horses towing an ice-boat on the Grand Union Canal in the early 1930s, the men holding on to a centre rail and rolling the craft to break the ice

SHIP CANALS: (*above*) a ship passes through the newly-opened Panama Canal about 1914; (*below*) the Trollhätte locks in Sweden in 1861. The flight shown has itself now been replaced by locks able to take 2,000-ton ships

PANAMA CANAL: (*above*) the French company at work on the Culebra Cut of the sea-level canal in 1888; (*below*) a barge-carrying ship passes a lock in 1970

EUROPEAN SCENES: (*above*) Briare aqueduct, designed by Eiffel, on the Canal Latéral de la Loire in France; (*below*) barges on the Danube at Passau, Germany

AMERICAN SCENES: (*above*) peace: a mule-towed barge, with a covered bridge behind, on the Delaware Division of the Pennsylvania Canal at Narrowsville; (*below*) war: barges piled with refugees' belongings at Richmond, Virginia, on the James River & Kanawha Canal during the Civil War

AMERICA: (*above*) a romantic picture of the future President Garfield, once a canal boy; (*below*) the Erie Canal at Lockport

AMERICAN BOATS: (*above*) a packet-boat prepares to leave Richmond for Lynchburg on the James River & Kanawha Canal in 1865; (*below*) drawing of two sections of a canal boat ascending one of the inclined planes on the Allegheny Portage Railroad section of the Pennsylvania main line

A boat cabin on an Erie Canal cargo boat

Marseilles. Or the Canal du Rh ontinent a-building from 1784, from Strasbou Rhin, also fifty years miles past a summit of 1,089 ft to the Saôn he Rhine for 201 interesting of them all is the Canal de la Mar But the most main French waterway going across the country Rhin, the om the

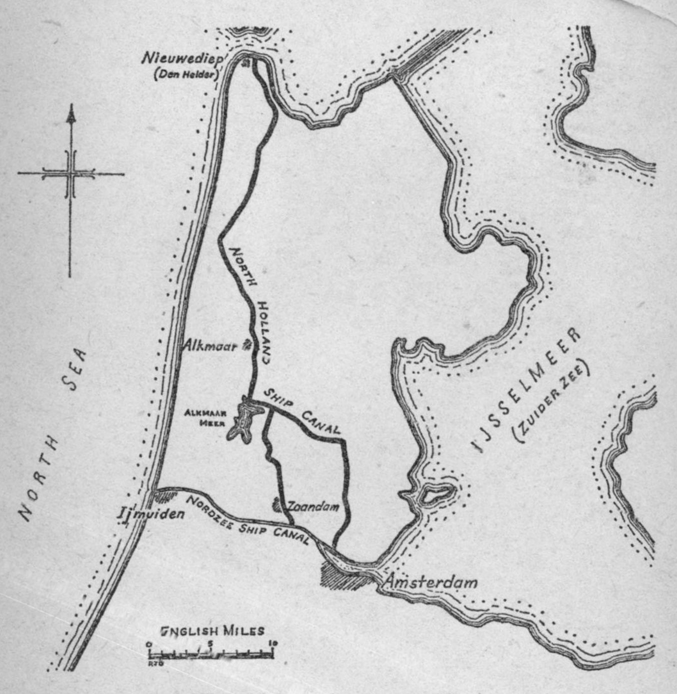

11 The ship canals of north Holland

Seine to the east. Built rather later than the other two, between 1838 and 1853, it starts at Vitry-le-François on the Marne lateral canal, with access to the Moselle, the canal line to Belgium, and the Seine. This astonishing waterway is 193½ miles long, running along narrow valleys to pass four mountain watersheds and a summit of 923 ft by its 178 locks on its wa past Nancy to S…g. Among its works are five tunnels, the longest, Ma…s, over 2½ miles, and the new Saint-Louis–Arzviller in… plane, first operated in 1968, which replaces 18 locks. …hose who explain that Continental canals run throu… flat country, and therefore have a natural advantage

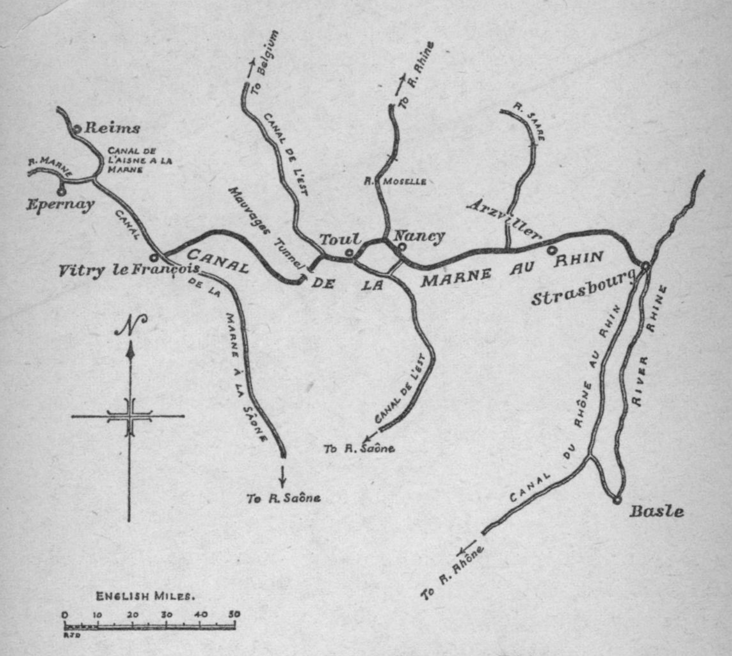

12 The Canal de la Marne au Rhin

over those in Britain (where the highest point on a canal still in use is 518 ft) might look at the Canals de Bourgogne, du Rhône au Rhin, and de la Marne au Rhin.

An important influence of the early nineteenth century on water transport was the steam boat. One of the first in Continental Europe had been a small paddle-wheeler, financed and built by two Americans then in France, Robert Fulton and the …mbassador, Robert R. Livingston. With two craft in tow, it was tested on the Seine at Paris in August 1803. The *Journal des Debats* was farsighted:

> This mechanism applied to our rivers, the Seine, the Loire and the Rhône, would be fraught with the most advantageous consequences to our internal navigation. The tows of barges which now require four months to come from Nantes to Paris would arrive promptly in 10 to 15 days.

In 1816 British-built craft started work on the Rhine and the Elbe; in 1830 steamers were working on the upper Danube, and in 1844 on the Brussels–Charleroi Canal.

Steam tugs or steam ships became almost universal on the big rivers; but most of the canals stayed faithful to the horse, until the coming of the now nearly universal diesel. I still treasure the memory of seeing the last of the Rhine steam paddle-wheel tugs pulling her barges up towards the Lorelei rock in 1963. But when in 1969 I explored the Dunube from Passav down to the great new Irm Gates locks, I found three of them still hard at work.

In parts of France, however, electric towing from the canal bank was introduced in 1895, and spread rapidly in the north and east. There were two systems. One used tractors drawing current from overhead wires, the other locomotives doing the same, but running on metre-gauge tracks. As the track was single, these could not pass, but when they met, they exchanged towlines, and returned the way they had come. By 1935 over 1,000 miles of canal were thus worked, but after that the system contracted, serving only older barges that were self-propelled, until it ended in 1968.

Greater towing power against a river current, without erosion of canal banks by the wash from propellers or paddles, could be obtained by the cumbersome system of laying a chain or wire cable on the bottom, and using a tug that hauled itself and a tow of barges along by picking up the chain on a steam-driven drum and then dropping it back again. Where a tug going one way met another, one had hurriedly to drop the chain and then grapple it again. This system was hardly used in Britain, except by the tug put on in 1826 to tow barges through the Islington tunnel in London, but in the New World it appeared on part of the St Lawrence and on 42 miles at the Buffalo end of the Erie Canal.

The first successful trial of chain towing on the Continent was made by M. de Rigny at Rouen; it led to systems, mostly installed in the 1860s and 1870s, on the Seine and Meuse and on some lengths of French and Belgian canals; on the Neckar and the Rhine from Bingen to Rotterdam; a continuous stretch of 425 miles of the Elbe; the Danube; and on 210 miles of the Volga. Some of them lasted well into our own century. Chain tugs on the big rivers were astonishingly big; those on the Elbe were 138 ft to 150 ft long, and 24 ft wide, but drawing only 18 ins of water, towing up to eight barges of 150 to 400 tons.

Robert Louis Stevenson saw such a barge train on the Willebroek Canal between the Schelde and Brussels in the 1870s:

> Every now and then we met or overtook a long string of boats, with great green tillers; high sterns with a window on either side of the rudder, and perhaps a jug or a flower-pot in one of the windows; a dinghy following behind; a woman busied about the day's dinner, and a handful of children. These barges were all tied one behind the other, with two ropes, to the number of twenty-five or thirty; and the line was headed and kept in motion by a steamer of strange construction. It had neither paddle-wheel nor screw; but . . . it fetched up over its bow a small bright chain which lay along the bottom of the canal, and paying it out again over the stern, dragging itself forward, link by link, with its whole retinue of loaded skows.

The scene changes, but not completely; canals are modernized, but still the broad water stretches onwards past the poplars; the world moves faster, and so do the boats, but still the windows have their lace-curtains, the wife her budgerigar, the baby its playpen on deck, though now, up forward, are two or three motor bikes, or, on the big Rhine barges, even a small motor-car, ready for the owner's pleasure when he goes ashore.

R. L. Stevenson described a stoppage for lock repairs in the 1870s:

> Long lines of barges lay one after the other along the canal; many of them looking mighty spruce and shipshape in their jerkin of Archangel tar picked out in white and green. Some carried gay iron railings, and quite a parterre of flower-pots. Children played on the decks, as heedless of the rain as if they had been brought up on Loch Carron side; men fished over the gunwale, some of them under umbrellas; women did their washing; and every barge boasted its mongrel cur by way of watch-dog. Each one barked furiously at the canoes, running alongside until he had got to the end of his own ship, and so passing on the word to the dog aboard the next . . . these little cities by the canal side had a very odd effect upon the mind. They seemed, with their flower-pots and their smoking chimneys, their washings and dinners, a rooted piece of nature in the scene; and yet if only the canal below were to open, one junk after another would hoist sail or harness horses and swim away into all parts of France; and the impromptu hamlet would separate, house by house, to the four winds. The children who played together by the Sambre and Oise Canal, each at his father's threshold, when and where might they next meet?

Today, were such a stoppage to occur on one of the smaller waterways, the scene would not be very different. But of the great canals, the Albert, the Juliana, with their doubled, tripled and quadrupled locks, their broad, deep channels, day and night the diesels sound, the waves break from the bows, the wash plops against the piling, as the barges run at 10 or 12 knots on the business of the Common Market and the world. And on the rivers, you will see more traffic still, and maybe too the new push-towing, the barges lashed rigidly in front, the whole steered as a single unit by a powerful blunt-nosed, multiple-ruddered tug at the back.

Sit on the foredeck of one of the luxurious steamers of the Köln-Düsseldorfer company, and travel down the Rhine. From the great inland port of Basle, past the huge locks of the upper Rhine by-pass called the Canal d'Alsace, past Strasbourg, Karlsruhe, Mannheim, Mainz, Koblenz, Cologne and Düsseldorf, you will find the river traffic increasing all the way. From Mainz onwards there will never be a time when you cannot count a dozen barges in sight; from Bingen past the Lorelei rock to Koblenz you will wonder how your craft can safely find its way among the dozens jostling for position, shooting downwards, struggling upwards; till, past Düsseldorf, you will look to the right to Duisburg-Ruhrort, the greatest inland waterway port in the world; and so, still approached, followed, passed, by barges old, young, big, small, smart, snug, domestic, businesslike, you will come to Rotterdam, where the barges meet the ships. From the Grand Canal of China; from the Stecknitz, the Bereguardo, the Canal du Briare, du Midi; Ludwigs and the Göta; the Exeter, the Duke of Bridgewater's, the Trent & Mersey and the Grand Canal of Ireland, we come at last to Europoort. We are in the nineteen-seventies now; but look to your left, to the barge tiers, row on row, in Europe's newest port. The dominance of canals and of inland navigation as the only important means of moving heavy goods is over; but the canal age is still with us. The craft may change with the changing years, but the water remains, in river and cut alike, to serve the future as it has the past.

CHAPTER 14

The Age in North America

THE driving urge to build canals in North America arose in order to make possible the huge expansion westwards towards the area north and south of the Ohio river, the rapid development of the regions involved, and the transport back east of their vast quantity of products. For this, road transport was no solution. It was said in 1812 that a good team of 5 or 6 horses would take 18 to 35 days to carry 1 to 1½ tons of goods between Philadelphia and Pittsburgh. In the United States the direct east-west route was barred by the long range of the Allegheny mountains. The first lines to the west sought to pass round them, either from the sea-coast north to the St Lawrence, Lake Ontario and Lake Erie, or from the Gulf of Mexico up the Mississippi.

The sea-coast and navigable rivers were natural channels of trade; beyond them, roads were few and poor. Canals seemed the answer: to improve transport from inland places to tide-water; to by-pass falls on otherwise navigable rivers; to join rivers together, or in their own right to open up the interior. Americans knew how greatly waterways had helped to develop parts of Continental Europe; more closely, they had watched the quick growth and spectacular results of British canals built between 1760 and 1792. By that date, thirty American waterway companies had been incorporated, as the advance guard of many more, but in 1816 only about 100 miles of canal had been built in the United States. Then a rush began not dissimilar to the earlier canal mania in Britain. Some waterways never got finished; others were successful; most were useful. All were built under great natural difficulties, with a lack of engineers and an adequate labour force, over long distances, and always against problems of raising money. It was no mean feat for a relatively undeveloped country to achieve from its own resources.

The first canals, small compared to their successors, were mostly east of the Alleghenies: the South Hadley Falls Canal in Massachusetts and the Dismal Swamp Canal to link Chesapeake Bay near Norfolk with the Pasquotank river leading to the North Carolina Sounds, opened in 1794, though one also near New Orleans; the Santee Canal in 1800, 22 miles long with 12 locks to connect the Cooper and Santee rivers in South Carolina; and the Middlesex Canal in 1803, 27 miles long with 20 locks, to join the Charles river at Boston with the Merrimack and, it was hoped, the Connecticut. These were followed by a number of canals running to or near tidewater. Of these, the most important and profitable was the coal-carrying Delaware & Hudson, from Honesdale in north-eastern Pennsylvania to the Delaware river, and then from the Delaware at Port Jervis to the Hudson river at Rondout, whence there was access to New York. It was finished in 1828 for craft of 30 tons, and by 1853 had been enlarged to 140-ton standard.

But the greater drive was to and from the west. The oldest route was by the St Lawrence river to Lake Ontario, and then past Niagara Falls to Lake Erie. The St Lawrence had drawn men towards the Great Lakes ever since Jacques Cartier in 1536 had sailed up it for six weeks, hoping it would lead him to China, so that when he reached a series of great rapids, a little above what is now called Montreal, he called them Lachine. What was probably the first lock canal in North America, a little one for canoes built in 1779 at Cascade Point, was on the St Lawrence; two others, that at Coteau rapids with three locks, had been added when my great-grandfather passed through them in 1785. Later, timber rafts worked down the river, rapids and all, to be broken up at the end of the journey, the boatmen who had lived upon them in little huts returning for other rafts. Traffic grew rapidly on the navigable reaches between the falls, especially after steam boats arrived, but goods and passengers had to be carried round the impassable stretches from one boat to another. It was not until 1848 that a navigable waterway with a depth of 9 ft, built by the local governments, existed all the way from the sea to Lake Ontario. Today, rebuilt again to take ships carrying 25,000 tons of cargo, this route forms part of the St Lawrence Seaway.

But between Lakes Ontario and Erie interposed the 326 ft rise of Niagara Falls, round which a canal was also needed. It was first built by the private Welland Canal company, with 40 locks 110 ft × 22 ft and 8 ft deep, opened in 1829. Later bought by the Upper Canada Government, it was then rebuilt to conform to the dimensions of the St Lawrence route and reopened in 1845, now with 27 locks. Between 1913 and 1932, the Canadian Government built an almost new canal, with eight much larger locks, the present Welland, now also part of the Seaway. Onwards from Lake Erie there is a natural passage to Lake Huron, and thus to Lake Michigan, with Chicago on its southern shore. But Lake Superior was closed until the first Sault Ste Marie Canal and lock were built by the St Mary's Falls Ship Canal company, and opened in 1855. Three Sault (Soo) canals now join the lakes, two American and one Canadian, and carry a heavy traffic.

After the war of 1812, much Canadian and British effort was put into developing a safe route to the interior of the country and the Great Lakes, well away from the St Lawrence and the United States border. Above Montreal, the Ottawa river runs north-west from the St Lawrence. Three short canals had by 1834 been built past the Long Sault, Chute à Blandeau and Carillon rapids. From higher up the river also, Col John By of the Royal Engineers constructed the Rideau Canal for 124 miles past rivers and lakes to Kingston on Lake Ontario. It was opened in 1832, and with the Ottawa river was for a time the best steam boat navigation to the Lakes. The settlement that grew up where the Rideau Canal left the river was at first called Bytown, after the engineer. We know it now as Ottawa, the capital of the Canadian Federation.

The St Lawrence route did not suit Americans either, for it was mostly controlled by Canada, it ran near a frontier not yet free from war, and it provided only sea communication with the eastern seaboard of the United States. So the great Erie Canal was cut, 363 miles long with 82 locks, with a main line from Albany on the Hudson river above New York by the Mohawk river valley to Buffalo on Lake Erie, a 38-mile branch to Oswego on Lake Ontario, and other branches to Seneca Lake and elsewhere. Built by the State of New York, most of it through unsettled land, begun in 1817, opened in 1825, at the time the longest in the world, perhaps no other canal has ever been so phenomenally successful in fulfilling the classic idea of canal service. In one direction passed the produce of the corn-lands and stockyards of the west; in the other manufactured goods from eastern foundries and factories for the farmers and the fast-growing middle-western cities, and the immigrants to swell their numbers, voyaging five days on river and canal from New York to Buffalo. It had been a good investment, for it reinforced the already existing pre-eminence in foreign trade of New York over Baltimore, Boston and Philadelphia.

In twelve years its capital cost had been paid off, improvements made and branches constructed. By then, too, the improvement of the St Lawrence and Welland route had been stimulated. It was widened and deepened throughout as the traffic carried continued to grow. In 1883 tolls were abolished, and in 1918 it was replaced by a new waterway, the New York State Barge Canal, deeper, wider, with fewer locks, to take craft carrying 2,500 tons. Not now much used, this probably faces either enlargement or abandonment.

Even before its completion, the Erie had been joined by the Champlain Canal, completed in 1823, also by the State of New York, to Lake Champlain, though it was not until 1843 that there was access through to the St Lawrence by the Chambly Canal and the Richelieu river.

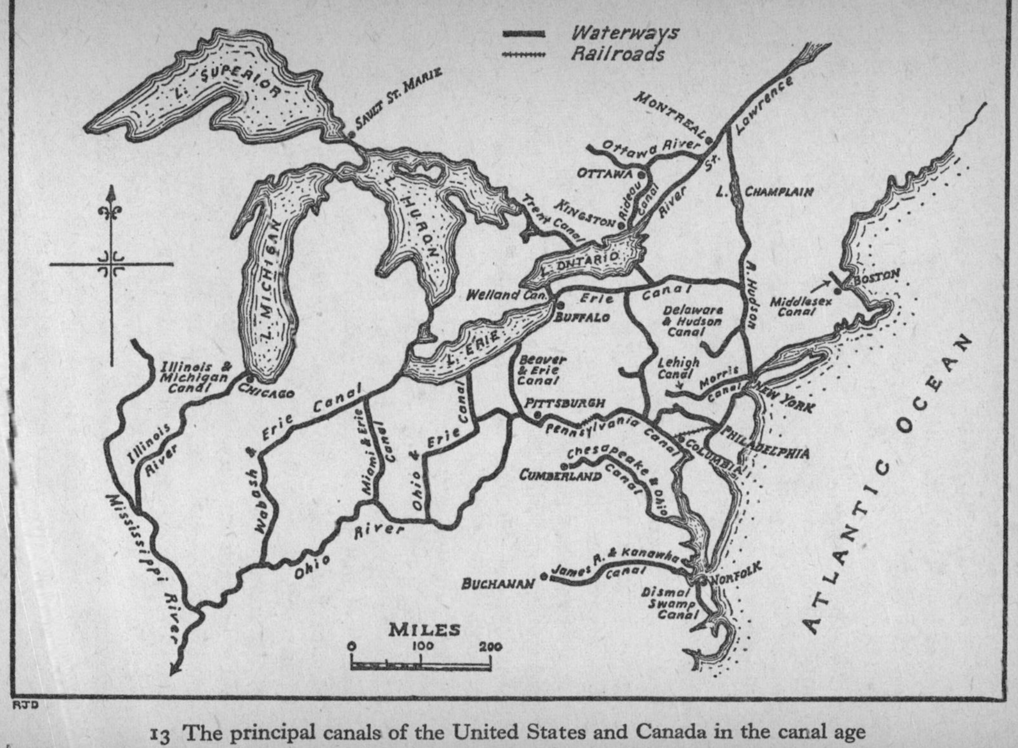

13 The principal canals of the United States and Canada in the canal age

To the south three ambitious efforts were made to link the eastern seaboard direct with the country beyond the Allegheny mountains. Two fell short, one was completed, the extraordinary State-financed and owned Pennsylvania Canal and Railroad, the 'Main Line' to link the Susquehanna to the Ohio river, 394 miles long between Philadelphia and Pittsburgh, opened in 1834.

A water route from Philadelphia had been opened in 1828 to Middletown on the Susquehanna river, 19 miles above Columbia, via the Schuylkill Navigation to Reading and the Union Canal to Middletown. But it was roundabout and heavily locked, and the locks on the Union Canal were only 8½ ft wide, against 17 ft on the Schuylkill. Therefore a railway was authorized in 1828, 82 miles long from Philadelphia direct to Columbia. It was opened in 1834 with an English-built locomotive, *Black Hawk*, and included two steam-engine operated inclined planes, one of 187 ft rise near Philadelphia and another of 90 ft at Columbia.

Then followed 170 miles of canal and river navigation, with 102 locks, 17 ft and 15 ft wide, past Middletown and up the valleys of the Susquehanna and Juniata to Hollidaysburg. A feature of this section was the level crossing of the Susquehanna at Duncan's Island made by building a dam across the river nearly 2,000 ft long, with a towing-path bridge a little above, from which the barges could be hauled in the still water between the dam and the bridge. In England similar level crossings took the Derby Canal across the Derwent in Derby, and the Uttoxeter branch of the Trent & Mersey over the Churnet. Just above this crossing, a 600-ft wooden trough aqueduct, with towing path on one side and path for foot passengers on the other, carried the canal over the Juniata.

Arrived at Hollidaysburg, passengers and freight changed to the 37 miles of the Allegheny Portage Railroad, which rose 1,399 ft by five inclined planes to the summit station and inn at Lemon House at 2,334 ft, and then fell 1,150 ft to Johnstown on the far side of the mountains. Traffic on the planes was counter-balanced when possible, stationary steam-engines being used to give what extra power was necessary. The longest of the planes was 3,100 ft, and the steepest 1 in 10. Some stretches between the planes were worked upwards by horses, others by locomotives, and downwards by gravity. 'Peregrin Prolix', an author from Philadelphia, having reached the summit, wrote:

> The ascending apprehension has left you, but it is succeeded by the fear of the steep descent which lies before you and as the car rolls along on this giddy height the thought trembled in your mind that it may slip over the head of the first descending plane, rush down the frightful steep and be dashed into a thousand pieces.

Charles Dickens was calmer:

> Occasionally the rails are laid upon the extreme verge of a giddy precipice and looking from the carriage window, the traveller gazes sheer down, without a stone or scrap of fence between into the mountain depths below. The journey is very carefully made, however, only two carriages travelling together and while proper precautions are taken, is not to be dreaded for its dangers.

At Johnstown another canal, 105 miles long with 68 locks, ran to Pittsburgh, which it entered by a long wooden-trough aqueduct 1,140 ft long over the Allegheny river (Dickens called it 'a vast low, wooden chamber full of water'), with a branch through a 810-ft tunnel to the Monongahela river, navigable then and now.

At first all passengers and freight had to be transhipped from railway to canal and back to railway and canal again, but later some of each worked through from Philadelphia to Pittsburgh in canal boats. These, built in three or four watertight sections, each some 20 ft long, could be separated for carriage on the railways, or be joined together by iron clamps for the waterway part of the route, after the railway trucks carrying them had been run down inclines into the water and the boat sections floated off.

In 1852 the Pennsylvania Railroad had paralleled the two canal lines to Hollidaysburg and between Johnstown and Pittsburgh, to connect with the Portage Railroad, and two years later it opened its own line over the Allegheny Mountains. In 1857 the railroad company bought the old canal and rail line. The Portage Railway was soon closed, the Johnstown–Pittsburgh canal section soon afterwards, but part of that to the east survived to the end of the century. One year earlier, in 1856, the Pennsylvania Canal system had at last been linked to the Erie Canal by a series of canals up the Susquehanna valley to the Chemung Canal and Seneca Lake branch of the Erie, and for a time enabled Pennsylvania coal to reach markets along the Erie Canal.

South again, the Ohio company had been formed in 1749 to exploit the fur trade of the west, and to settle new lands. It pushed up the valley of the Potomac, using the river to Cumberland and land transport round the falls, and then a track over the mountains to the Monongahela river. But it declined until the enthusiasm of George Washington, who had once been a land surveyor, caused the formation of the Potomac Company in 1785, to make the river navigable at least to Cumberland on the way to the Ohio. Struggling against the difficulties of men and nature, the company built canals round the rapids, but could do little to improve the main river. The by-pass round Great Falls was an extraordinary engineering feat, for the locks, rising 76 ft in ⅜ mile, had to be cut partly in the solid rock; it was similar in many ways to the problem faced by those who built the Trollhätte Canal out of the Göta river in Sweden. However, by 1802 the Potomac was navigable after a fashion for about 220 miles, and the company then turned to making some of the tributaries passable, notably the Shenandoah. But money was hard to raise and the rivers were too rough. By 1822 it was moribund, and it died in 1828.

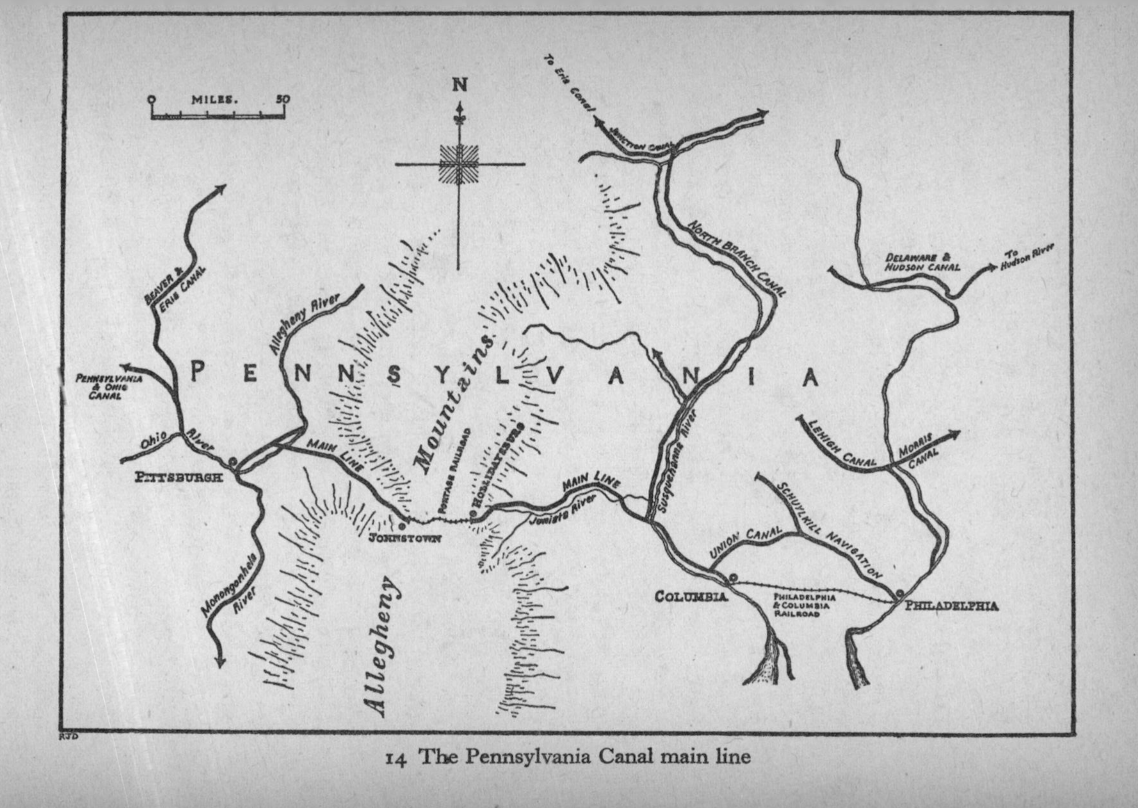

14 The Pennsylvania Canal main line

By then the Erie Canal was showing what canals could do, and the Pennsylvania had been begun. A canal was clearly also the heavy freight answer to the troublesome Potomac and the route to the west, even though the National Road had been finished in 1820 from Baltimore to Wheeling on the Ohio in west Virginia. Work began in 1828 on the Chesapeake & Ohio Canal, intended to leave tidewater near the city of Washington, and to take its still-water line past Cumberland and over the mountains to Pittsburgh. In 1829 the company were advertising in England for one or two thousand men, stonemasons and others, to be interviewed by Mr Richards, their agent at Liverpool, for work on the canal – in those days a muscle drain. But, as the Potomac company had found, money, mainly from public sources, was difficult to get, and by 1842 only 135 miles had been built, the end still nearly 50 miles short of the coal-fields at Cumberland, which the rival Baltimore & Ohio Railway had reached in that same year. The waterway then mainly carried the agricultural produce of the neighbourhood and a single daily packet boat running toll-free.

By 1850 it had indeed reached Cumberland, and was now 179½ miles long with 75 locks. It got no further, but coal and other traffic grew steadily, till the 204,000 tons of 1851 became the 974,000 of 1875. Then came the depression of 1876, rate cutting by the railway, a serious strike, and a Potomac flood that greatly damaged the canal. It was restored, and again wrecked by a flood in 1886. It never fully recovered, and passed into the railway's hands. Its working life ended with still another flood in 1924; all that now remains is a section at Washington used as a national park.

As George Washington had been the father of the Potomac company, so he was of the improvement of the James river that runs from Chesapeake Bay past Richmond, Virginia. The James River Company was formed in 1784, its object to improve the river to a point near what is now Buchanan. The company built a 7-mile canal round the falls at Richmond, and somewhat improved the river navigation. Commercially it was successful, the corn, tobacco and coal trade enabling it to pay 12 per cent, but the navigation mostly remained elementary. In 1812 an investigating committee proposed making the upper James navigable, building a road over the mountains, and making the Greenbrier river on the other side navigable to the great falls of Kanawha, beyond which the Kanawha river ran to the Ohio. But the company had not the money to start. In 1820 the State took it over and, altering the committee's plans, during the next ten years built a 200-mile long mountain road from Covington on the upper James to the Ohio, and improved the river navigation. They also started on a canal upwards of Richmond towards the Covington road. This last effort did not get far, and in 1832 was transferred to the James River & Kanawha company, which planned to continue the canal upwards, and then to connect it by railway with the Ohio. The company was strongly supported by the city of Richmond, which hoped to share in the rapidly growing trade with the west that New York, Philadelphia and Baltimore were also seeking by means of their canals.

The James River & Kanawha Canal was completed to Lynchburg in 1840 and Buchanan in 1851. It consisted of 160 miles of canal and 37 of river, able to take 60–80-ton barges and the usual packet boats, with 90 locks rising 728 ft. Some work, including tunnelling, was done beyond Buchanan towards Covington, and in the early 1850s surveys and plans were made for extending the water line over the Alleghenies, 480 miles in all from Richmond to the junction of the Kanawha with the Ohio, one plan including a 9-mile tunnel 44 ft wide and 32 ft high with a double towpath. But the canal was in Confederate territory, and the Civil War ruined it. Afterwards, in spite of some revival and peak traffic of 240,000 tons in 1860, and of mid-west interest in a water route from the Ohio alternative to railways to keep rates down, it declined until the disastrous flood of 1877. The following year the company sold out to the Richmond & Allegheny Railway, the track of which was laid partly on the towpath.

All these routes had run from east to west; that to the south was the Mississippi itself, with its great tributary the Ohio, and the other rivers that joined it. The Louisiana Purchase of 1803 transferred control from France to the United States, after which much of the produce of the mid-west was carried down its muddy waters to New Orleans and the Gulf of Mexico. Steam boats came, and navigation works were begun. One was the 2-mile long Louisville & Portland Canal round falls of the Ohio, opened in 1830 with three of the biggest locks so far built in the United States, 183 ft × 50 ft, though these were soon to prove too small for the river steam boats' increasing size. Another was the 96-mile long Illinois–Michigan Canal, a cut from the Illinois river, a tributary of the Mississippi, to Lake Michigan. Chicago, with fewer than 200 inhabitants, was not yet a town when the canal commissioners arrived to mark out the point of junction with the lake; the survey they filed in 1830 is the beginning of Chicago's municipal history. The waterway was opened fifteen years later, and contributed greatly to the city's commercial growth. It was replaced in 1900 by the present ship canal and enlarged river route to the Mississippi.

From the Ohio river there was a drive to the north, to the Lakes, to the Erie Canal, the Welland and the St Lawrence. So a number of canals were built from the river to Lake Erie. Farthest to the east was the Beaver & Erie, opened in 1844 as a continuation of the Pennsylvania Canal main line to Pittsburgh. From the Ohio at the Beaver river below Pittsburgh it ran 136 miles by canal and river to Erie, with 137 locks. Next to the west was the Ohio & Erie, from Portsmouth on the Ohio by way of the Scioto and other rivers to Cleveland on the lake past Colombus, Newark, New Philadelphia and Akron, 307 miles with 150 locks, begun in 1825 and completed in 1833. Its traffic reached a peak in 1851, though it lasted until 1913. This canal had a second link with the Ohio at Marietta and a third by way of the Pennsylvania & Erie or Cross-Cut Canal, which joined it to the Beaver & Erie. It was on the Ohio &

Erie that a future American President, James Garfield, worked as a canal boy for four months in 1847.

Westwards, the Miami & Erie, 244 miles and 93 locks from Cincinnati on the Ohio by way of the Great Miami river to Toledo on the lake, was opened to Dayton in 1832, but through to Toledo not until 1845. Part of its course on the Miami to Toledo was the same as that of the fifth, the Wabash & Erie, over 450 miles long from Evansville on the Ohio, begun in 1832 but not completed throughout until 1853. Its lower portion had an active life of only nine years.

Tremendous energy was put into building the Ohio canals, as into those in Pennsylvania, New York and elsewhere, until the same energy was transferred to railroad construction. One is astonished by this drive and persistence. So was Mrs Trollope, when she saw the 'extraordinary work' of the Morris Canal between the Hudson and Delaware rivers, which

> at one point runs along the side of a mountain at thirty feet above the tops of the highest buildings in the town of Paterson, below; at another, crosses the falls of the Passaic, in a stone aqueduct, sixty feet above the water.

She added, justly:

> There is no point in the national character of the Americans which commands so much respect as the boldness and energy with which public works are undertaken and carried through. Nothing stops them if a profitable result can be fairly hoped for.

The Morris, with the Lehigh Canal, was a busy coal-carrying waterway. The source was Lehigh anthracite mining district in eastern Pennsylvania; the markets were Philadelphia, New York, and along the Susquehanna; the driving force a Quaker industrialist, Josiah White; the means the Lehigh Coal & Navigation Company, founded in 1822 and still in existence. In the same decade he completed the Lehigh Canal down that river to Easton, whence in 1832 there was access to Philadelphia down the Delaware Division of the Pennsylvania Canal – and, incidentally, also built the first railway in the United States, a 9-mile long gravity-operated coal line from Summit Hill. By 1860 the Lehigh was carrying well over a million tons a year of traffic in barges up to 100 tons.

From Easton the Morris Canal carried his coal in sectional boats for 102 miles past its 23 inclined planes and 23 locks to Jersey City on the Hudson river, and so to New York. By 1866 the Morris Canal was carrying 889,220 tons, of which 473,028 were coal. As for Susquehanna markets, Josiah White extended his Lehigh waterway farther up the Lehigh, using very big locks, 100 ft long, 20 ft wide, with lifts up to 30 ft, and then by a 25-mile railway to Wilkes-Barre on the North Branch of the Susquehanna, past three inclined planes and a 600-yd tunnel.

By 1850, over 3,500 miles of canal had been built in the United States. Yet canals, in the sense of wholly artificial waterways, nevertheless played a minor part in early American transport – with the one exception of the Erie – but inland navigation did not. A reader of Charles Dickens' *American Notes* will see how much of his travelling in 1842 is by steam boat, on the sea, lakes, rivers or canals, and how little by road or rail. On the second part of his adventures in North America, here is how he travelled: Baltimore–Harrisburg, rail and road; Harrisburg–Pittsburgh, by the Pennsylvania Canal and Portage railroad; Pittsburgh to St Louis, down the Ohio river and up the Mississippi, by steam boat, and so back again to Cincinnati on the Ohio; road and rail to Lake Erie; then almost always by boat via Niagara, Lake Ontario and the St Lawrence to Montreal; rail, boat on Lake Champlain, and road to Albany, ending with a boat down the Hudson river to New York. The extent of natural rivers and of the Great Lakes, the difficulty of finding money to build roads and railways in a huge, sparsely populated country with rapidly expanding frontiers, caused water transport to play an exceptionally large part in development.

Today, almost all the canals described, except the Seaway and the New York State Barge Canal, have gone, their moss-grown locks of interest only to romantic antiquarians and national park authorities. But the Mississippi runs for 1,837 miles from Minneapolis to New Orleans, its tug boats pushing groups of barges that may take 20,000 tons or more, carrying a great trade gathered also from the Arkansas, the Missouri, the Ohio, the Illinois which leads by the Illinois & Michigan Canal to Lake Michigan, and many others. At New Orleans it joins the Gulf Intracoastal Waterway, a mixture of canals and natural channels that runs from the Mexican border to Apalachee Bay, Florida. The cross-Florida canal is now being built by US Army engineers, 185 miles long and 12 ft deep from Yankeetown to Jacksonville to meet the Atlantic Intracoastal, 1,000 miles upwards to Norfolk, Virginia. This takes in several canals, including the Chesapeake & Delaware. One is the Dismal Swamp, which thus links the beginning of American canal building to the present day.

These and many other waterways are also used by tens of thousands of pleasure craft, on a scale not seen in Europe. In Canada, too, pleasure cruising is popular, on lakes and rivers, on the Rideau, and on the Trent Canal, a system of lakes, rivers and canals that leads from Trenton, Lake Ontario, to Georgian Bay, Lake Huron, 241 miles long with 42 locks, two vertical lifts and two small marine railways for transporting pleasure boats.

As in Europe, waterways decline and are built, uses change, craft develop, but the age continues.

APPENDIX I

Growth of the Inland Navigation System in England and Wales

Year	*Miles*
1760	1,398¼
1770	1,617⅛
1780	2,091⅜
1790	2,223¼
1800	3,074¾
1810	3,456⅜
1820	3,691¼
1830	3,875½
1840	4,003
1850	4,023⅜

For the Midlands, South and Wales I have used the figures given in my books, *The Canals of Southern England*, *The Canals of South Wales and the Border*, *The Canals of the West Midlands* and *The Canals of the East Midlands*. For the North and East I have made the best estimates possible in the present state of knowledge.

APPENDIX II

Inland Navigation and the Growth of Towns, 1801-41

Towns in England and Wales, on inland waterways only, with the dates (if 1760 or later) when they were given water communication, and their populations in 1801 and 1841. (Based on Mitchell, B. R., *Abstract of British Historical Statistics*, 1962: Population and Vital Statistics, Table 8, p 24.)

Town	*Date given inland navigation*	*Population* 1801	1841
		000s	
Bath	Before 1760[1]	33	53
Birmingham	1772[2]	71	183
Blackburn	1810	12	37
Bolton	1796	18	51
Bradford	1774	13	67
Cambridge	Before 1760	10	24
Carlisle	1823	9	22
Coventry	1769[3]	16	31
Derby	1795	11	33
Dudley	1779	10	31
Exeter	Before 1760	17	31
Halifax	*c* 1770[4]	12	28
Huddersfield	*c* 1777[5]	7	25
Leeds	Before 1760[6]	53	152
Leicester	1794[7]	17	53
Macclesfield	1831	11	33
Manchester	Before 1760[8]	75	235
Northampton	1761[9]	7	21
Norwich	Before 1760	36	62
Nottingham	Before 1760[10]	29	52
Oldham	*c* 1796	12	43
Oxford	Before 1760[11]	12	24
Reading	Before 1760[12]	10	19
Salford	Before 1760[13]	14	53
Sheffield	1819[14]	46	111
Shrewsbury	Before 1760[15]	15	18
Stockport	1797	17	50

Town	*Date given inland navigation*	*Population* 1801	1841
		000*s*	
Stoke-on-Trent	1772	28 (1811)	54
Wakefield	Before 1760[16]	11	19
Walsall	*c* 1798	10	20
Wigan	1774[17]	11	26
Wolverhampton	1770[18]	13	36
Worcester	Before 1760[19]	11	27
York	Before 1760	17	29

[1] By the Bristol Avon. Connected to the Thames by the Kennet & Avon Canal, 1810.

[2] Date of the first canal to Birmingham. Many others were built later.

[3] Connected to coalfield. Joined to main canal system, 1790.

[4] Date of the Calder & Hebble, near Halifax. The branch into the town was opened in 1828.

[5] Date of Sir John Ramsden's canal branch from the Calder & Hebble. Connected to Manchester by the Huddersfield Canal, 1811.

[6] By the Aire & Calder. Connected by water to Liverpool by the Leeds & Liverpool Canal, 1816.

[7] From the Trent. Connected through to London by water, by the Grand Union Canal, 1814.

[8] By the Mersey & Irwell to Liverpool. Many other canals were built later.

[9] By the Nene. Connected to the main canal system, 1815.

[10] By the Trent. Nottingham Canal to the coalfields opened 1796.

[11] By the Thames. Oxford Canal to the Midlands opened 1790.

[12] By the Thames and Kennet. Connected to Bristol by the Kennet & Avon Canal, 1810.

[13] By the Mersey & Irwell to Liverpool. Other canals built later.

[14] By the Don to Tinsley. Canal from Tinsley to Sheffield opened 1819.

[15] By the Severn. Shrewsbury Canal opened to the coalfields, 1796. Connected to the main canal system, 1835.

[16] By the Aire & Calder. The Calder & Hebble from Wakefield to Sowerby Bridge was opened *c* 1770. Canal communication to Manchester by the Rochdale Canal, 1804.

[17] By a section of the Douglas Navigation, and the Leeds & Liverpool Canal, later connected to Liverpool by canal throughout. Linked to Manchester, 1821.

[18] Date of the opening of the Staffs & Worcs Canal to Compton near Wolverhampton. The Birmingham Canal through the town to join the Staffs and Worcs was opened 1772.

[19] By the Severn. Connected to Birmingham by the Staffs & Worcs and Birmingham canals, 1772.

APPENDIX III

Construction Cost Per Mile of Some Canals at Different Periods

I. *Canals completed by 1790*

Waterway[1]	*Date of Act*	*Date completed*	*Index*[2]	*Cost per mile* £
Trent & Mersey (*N*)	1766	1777	99–119	3,213
Staffs & Worcs (*N*)	1766	1772	99–117	2,168
Birmingham (*N*)	1768	1772	99–117	4,091
Coventry (*N*)	1768	1790	99–129	5,395
Oxford (*N*)	1769	1790	99–129	3,374
Chesterfield (*N*)	1771	1777	107–119	3,231
Erewash (*B*)	1777	1779	108–117	1,787

Average figure £3,323

II. *Canals started and finished in the 1790s*

Waterway[1]	*Date of Act*	*Date completed*	*Index*[2]	*Cost per mile* £
Glamorganshire (*N*)	1790	1794	124–136	4,063
Nottingham (*B*)	1792	1796	122–154	5,246
Shrewsbury (*N*)	1793	1796	129–154	4,118
Grantham (*B*)	1793	1797	129–154	3,591

Average figure £4,256

III. *Canals begun in the 1790s, but finished after 1799 and before 1816*

Waterway[1]	*Date of Act*	*Date completed*	*Index*[2]	*Cost per mile* £
Worcester & Birmingham (*N*)	1791	1815	121–243	20,333
Brecknock & Abergavenny (*N*)	1793	1812	129–237	6,015
Grand Junction (*B*)	1793	1805	129–228	13,221
Oakham (*B*)	1793	1803	129–228	4,590
Warwick & Birmingham (*N*)	1793	1800	129–212	7,072
Ashby-de-la-Zouch (*B*)	1794	1804	136–228	5,543
Kennet & Avon (*B*)	1794	1810	136–228	16,666
Wilts & Berks (*N*)	1796	1810	154–228	4,357

Average figure £9,725

IV. *Canals begun after 1800*

Waterway[1]	*Date of Act*	*Date completed*	*Index*[2]	*Cost per mile* £
Croydon (*B*)	1801	1809	212–228	13,730
Grand Union (*N*)	1810	1814	206–243	12,281
Montgomeryshire (Western Branch) (*N*)	1815	1821	139–194	7,241
Macclesfield (*N*)	1826	1831	[3]	12,249
Birmingham & Liverpool Junction (*N*)	1826	1835	[3]	16,000
Stourbridge Extension (*N*)	1837	1840	[3]	16,333

Average figure £12,972

[1] (*N*) or (*B*) after the name denotes Narrow or Broad canal.

[2] In this column is given the range of the consumer price index figures for the construction period of each canal. I have used the Schumpeter-Gilboy series.

[3] Outside the price index used.

APPENDIX IV

Dividends of Some Canal Companies in 1830[1]

I. *Canals completed by 1790*

Waterway	*Dividend per cent*
Trent & Mersey	75 (probable)
Staffs & Worcs	25·7
Birmingham	100 (after allowing for bonus shares)
Coventry	49
Oxford	32
Chesterfield	7 (probable)
Erewash	48
	Average rate 48·1 per cent

The Chesterfield Canal, an isolated branch off the tidal Trent, is the only canal in this group not to have been linked to others built later.

II. *Canals started and finished in the 1790s*

Waterway	*Dividend per cent*
Glamorganshire	8[2]
Nottingham	12
Shrewsbury	8·8
Grantham	6·66
	Average rate 8·87 per cent, or 13·12 if the Glamorganshire is taken at 25 per cent

[1] As the financial years of companies were not always the same, the dividends quoted are for slightly different periods.

[2] The rate of dividend of the Glamorganshire was limited by its Act to 8 per cent, surpluses going to reduce tolls to an abnormally low level. It seems likely that without the limitation, and with rather higher tolls, about 25 per cent could have been paid.

III. *Canals begun in the 1790s, but finished after 1799 and before 1816*

Waterway	*Dividend per cent*
Worcester & Birmingham	4·45
Brecknock & Abergavenny	3·33
Grand Junction	13
Oakham	1·54
Warwick & Birmingham	12·5
Ashby-de-la-Zouch	3·54
Kennet & Avon	3·125
Wilts & Berks	none
	Average rate 5·186 per cent

The Grand Junction and the Warwick & Birmingham both formed part of an all-canal direct line from Birmingham to London that from 1805 replaced the previous circuitous canal and river route via Fazeley, Coventry, Oxford and the Thames. They therefore did exceptionally well.

IV. *Canals begun after 1800*

Croydon	1 estimated
Grand Union	none
Montgomeryshire (Western Branch)	none
Macclesfield	unfinished in 1830[1]
Birmingham & Liverpool Junction	unfinished in 1830[2]
Stourbridge Extension	not yet begun[3]

The last three canals in this group were opened in 1831, 1835 and 1840 respectively. No canal in this group paid more than the 4½ per cent of the Stourbridge Extension before they were all bought by railways, or absorbed by railway-controlled companies, in 1845 and the years immediately following.

[1] Highest dividend before bought by railway in 1846, 2·5 per cent during 1836 to 1838.

[2] Paid no dividend before amalgamation in 1845.

[3] Highest dividend before bought by railway company in 1846, 4·5 per cent for 1845.

Bibliography

I have first listed the principal printed sources I have used in writing each chapter of *The Canal Age*, though by far the greater part of its contents is derived from manuscript material such as minute books, most of which will be found in British Transport Historical Records, 66 Porchester Road, London, W2. Later I have listed such other general books as may be useful to the reader. A longer bibliography will be found in the fourth edition of my *British Canals*, and details of specialized sources of information about individual canals in the Notes at the back of each volume of the 'Canals of the British Isles' series of regional histories.

Chapter 1

Needham, Joseph, 'China and the invention of the Pound-Lock', *Procs Newcomen Soc*, *1964*.

Skempton, Prof A. W., 'Canals and River Navigations before 1750', *A History of Technology*, III, 1957.

Chapter 2

Barker, T. C. & Harris, J. R., *A Merseyside Town in the Industrial Revolution: St Helens 1750–1900*, 1954.

Clegg, Herbert, 'The Third Duke of Bridgewater's Canal Works in Manchester', *Trans Lancs & Cheshire Antiq Soc*, LXV, 1955.

McCutcheon, W. A., *The Canals of the North of Ireland*, 1965.

Malet, Hugh, *The Canal Duke*, 1961.

Mullineux, Frank, 'The Duke of Bridgewater's Underground Canals at Worsley', *Trans Lancs & Cheshire Antiq Soc*, LXXI, 1961.

Skempton, Prof A. W., 'The Engineers of the English River Navigations, 1620–1760', *Procs Newcomen Soc*, 1953.

Tomlinson, V. I., 'Early Warehouses on Manchester Waterways', *Trans Lancs & Cheshire Antiq Soc*, LXXI, 1961.

Tomlinson, V. I., 'Salford Activities connected with the Bridgewater Navigation', *Trans Lancs & Cheshire Antiq Soc*, LXVI, 1956.

Chapter 3

King-Hele, Desmond, *Erasmus Darwin, 1731–1802*, 1963.

Raistrick, Arthur, *Dynasty of Ironfounders. The Darbys and Coalbrookdale*, 1953.

Schofield, Robert E., *The Lunar Society of Birmingham*, 1963.

The Wedgwood correspondence (privately printed).

Chapter 4

Pressnell, L. S., *Country Banking in the Industrial Revolution*, 1956.

Chapter 5

Rees, A., *Cyclopaedia*, art 'Canal', 1805.

Chapter 6

Allnutt, Z., *Useful and Correct Accounts of the Navigation of the Rivers and Canals west of London* . . . ND.

Baxter, B., *Stone Blocks and Iron Rails* (*Tramroads*), 1966.

Copeland, John, *The Roads and their Traffic, 1750–1850*, 1968.

Halfpenny, E., '"Pickfords". Expansion and Crisis in the early nineteenth century', *Business History*, June 1959.

Chapter 7

Anon, *Our Cruise in the Undine*, 1854.

Anon, *The Waterway to London, as explored in the 'Wanderer' and 'Ranger', with Sail, Paddle and Oar, in a voyage on the Mersey, Perry, Severn and several canals*, 1869.

Anon, *What may be Done in Two Months. A Summer's Tour through Belgium, up the Rhine*, etc, 1834.

Aubertin, C. J., *A Caravan Afloat*, ND.

British Waterways: Recreation and Amenity, HMSO, Cmd 3401, 1967.
'Bumps', *A Trip through the Caledonian Canal*, 1861 (privately printed).
ed Coombs, H. & Bax, A. N., *Journal of a Somerset Rector*, 1930.
Cooper, T. Sidney, *My Life*, new ed 1891.
Dashwood, J. B., *The Thames to the Solent by canal and sea, or the Log of the Una Boat 'Caprice'*, 1868.
Pilkington, Roger, *Small Boat in Southern France*, 1965.
'Red Rover', *Canal and River: a canoe cruise from Leicestershire to Greenhithe*, 1873.
Karl Bernhard, Duke of Saxe-Weimar Eisenach, *Travels through North America, During the Years 1825 and 1826*, I.
Scenes and Songs of the Erie–Ohio Canal, Ohio State Arch & Hist Soc, 1952.
Southey, R., *Journal of a Tour in the Netherlands in the Autumn of 1815*, 1903 ed.
Stevenson, R. L., *An Inland Voyage*, 1878.
Trollope, Anthony, *The Kellys and the O'Kellys*, 1848.
Trollope, Mrs Frances, *Domestic Manners of the Americans*, 1832.
Westall, George, *Inland Cruising on the Rivers and Canals of England and Wales*, 1908.

Chapter 9

The Canal Boatmen's Magazine, files, BM.
Pearse, Mark Guy, *Rob Rat, A Story of Barge Life*, ND.
Robins, William, *Paddington Past and Present*, 1853.

Chapter 10

Barker, T. C. & Harris, J. R., *A Merseyside Town in the Industrial Revolution: St Helens, 1750–1900*, 1954.
Copeland, John, *Roads and their Traffic, 1750–1850*, 1968.
Edwards, Michael M., *The Growth of the British Cotton Trade, 1780–1815*, 1967.

Chapter 12

Aitken, Hugh G. J., *The Welland Canal Company*, 1954.
Beatty, C., *Ferdinand de Lesseps*, 1956.
Chevrier, Lionel, *The St Lawrence Seaway*, 1959.
Cornish, Vaughan, *The Panama Canal and its Makers*, 1909.
De Lesseps, F., *The Suez Canal*, 1876.
Fitzgerald, Percy, *The Great Canal at Suez*, 1876.
Hadfield, Charles, *Canals of the World*, 1964.
Hammond, Rolt, and Lewin, C. J., *The Panama Canal*, 1966.
Howarth, David, *The Golden Isthmus*, 1966.
ed James, P., *The Travel Diaries of T. R. Malthus*, 1966.
Leech, Sir Bosdin, *History of the Manchester Ship Canal*, 1907.
The St Lawrence Seaway (Queen's Printer, Ottawa).
Siegfried, A., *Suez and Panama*, 1940.

Chapter 13

Calvert, Roger, *Inland Waterways of Europe*, 1963.
Dickinson, H. W., *Robert Fulton, Engineer and Artist*, 1913.
Hartley, Sir Charles A., 'Inland Navigations in Europe', *Procs ICE*, March 19th, 1885.
Jeans, J. S., *Waterways and Water Transport in different countries*, 1890.
Meyer, F. J. & Wernigh, W., *Steam Towing on Rivers and Canals by means of a submerged cable*, 1876.
Pilkington, Roger, 'Canals: Inland Waterways outside Britain', *The History of Technology*, IV, 1958.
Pilkington, Roger, *Small Boat in Southern France*, 1965.
Pilkington, Roger, *Small Boat through France*, 1964.
Skempton, Prof A. W., 'Canals and River Navigation before 1750', *A History of Technology*, III, 1957.
Stevenson, R. L., *An Inland Voyage*, 1878.

Chapter 14

The Canals of Canada (Queen's Printer, Ottawa).
Chalmers, Harvey, II, *The Birth of the Erie Canal*, 1960.

Dickens, Charles, *American Notes*, 1842.

Encyclopaedia Britannica, 7th ed, 1842, art 'Canal'.

Goodrich, Carter, *Canals and American Economic Development.*

Goodrich, Carter, *Government Promotion of American Canals and Railroads, 1800–1890*, 1960.

Harlow, Alvin F., *Old Towpaths*, 1926.

Kirkwood, James J., *Waterway to the West* (James River & Kanawha Canal), (Eastern National Park & Monument Association), 1963.

Legget, Robert, *Rideau Waterway*, 1955.

Phillpots, Lt Col, *Report on the Canal Navigation of the Canadas*, 1842.

Porcher, F. A., *The History of the Santee Canal* (South Carolina Public Service Authority), 1950.

Roberts, Christopher, *The Middlesex Canal, 1793–1860*, 1938.

ed Robertson, D. S., *An Englishman in America, 1785; the diary of Joseph Hadfield*, 1933.

Sanderlin, Walter S., *The Great National Project: a History of the Chesapeake & Ohio Canal*, 1946.

Shank, William H., *The Amazing Pennsylvania Canals* (The Historical Society of York County), 1965.

Taylor, George Rogers, *The Transportation Revolution, 1815–1860*, 1964.

Trollope, Mrs Frances, *Domestic Manners of the Americans*, 1832.

ed Walker, B. K. & W. S., *The Erie Canal, Gateway to Empire; Selected Source Materials*, 1963.

General

Anon, *The History of Inland Navigation*, 1766.

Calvert, Roy, *Inland Waterways of Europe*, 1963.

Delany, V. T. H. & D. R., *The Canals of the South of Ireland*, 1966.

De Salis, H. R., *Bradshaw's Canals and Navigable Rivers of England and Wales*, 1904.

Forbes, U. A. & Ashford, W. H. R., *Our Waterways. A history of inland navigation considered as a branch of water conservancy*, 1906.

Fulton, R., *A Treatise on the Improvement of Canal Navigation*, 1794.

Hadfield, Charles, *British Canals*, 4th ed., 1969.

Hadfield, Charles, *The Canals of South and South East England*, 1969.

Hadfield, Charles, *The Canals of South West England*, 1967.

Hadfield, Charles, *The Canals of South Wales & the Border*, 1967.

Hadfield, Charles, *The Canals of the West Midlands*, 1966.

Hadfield, Charles, *The Canals of the East Midlands*, 1966.

Hadfield Charles & Biddle, Gordon, *The Canals of North West England*, 1970.

Hadfield, Charles & Norris, John, *Waterways to Stratford*, 1968.

Jackman, W. T., *The Development of Transportation in Modern England*, 1914 (reprinted).

Lindsay, Jean, *The Canals of Scotland*, 1968.

McCutcheon, W. A., *The Canals of the North of Ireland*, 1965.

Phillips, J., *A General History of Inland Navigation*, 1792.

Pilkington, Roger, the 'Small Boat' series of books.

Priestley, J., *Historical Account of the Navigable Rivers, Canals and Railways throughout Great Britain*, 1831.

Rolt, L. T. C., *Thomas Telford: a Biography*, 1958.

Rolt, L. T. C., *The Inland Waterways of England*, 1950.

Royal Commission on the Canals and Inland Navigations of the United Kingdom, *Report*, etc, 12 vols, 1907–9.

Smiles, Samuel, *Lives of the Engineers*, 1861.

Smith, George, *Our Canal Population*, 1875.

Vernon-Harcourt, L. F., *Rivers and Canals*, 1896.

Index

Note: Subject entries are grouped under the word 'Canal'